Advancing Training and Inference Methods for Reinforcement Learning-Based Job Shop Scheduling

Constantin Waubert de Puiseau

Advancing Training and Inference Methods for Reinforcement Learning-Based Job Shop Scheduling

Springer Vieweg

Constantin Waubert de Puiseau
Bergische Universität Wuppertal
Wuppertal, Germany

The scientific studies and corresponding results presented in this book were conducted within the context of a doctoral project to obtain the degree of Dr.-Ing. at the School of Electrical, Information and Media Engineering at the University of Wuppertal.

ISBN 978-3-658-51304-7 ISBN 978-3-658-51305-4 (eBook)
https://doi.org/10.1007/978-3-658-51305-4

This Springer Vieweg imprint is published by the registered company Springer Fachmedien Wiesbaden GmbH, part of Springer Nature.
The registered company address is: Abraham-Lincoln-Str. 46, 65189 Wiesbaden, Germany

Abstract

Industrial production faces increasing operational challenges due to global supply chains, shorter development cycle times, and highly individualized products. At the same time, the growing availability of production data paired with advances in artificial intelligence (AI) can offer new optimization opportunities for complex production systems. Motivated by the high relevance of the arising challenges and AI as a promising remedy, this dissertation explores the application of deep reinforcement learning (DRL), a machine learning paradigm for sequential decision-making, to machine scheduling problems. The goal is to find DRL-based solutions that can generate better schedules in shorter computational times.

The thesis begins with a comprehensive literature review of 143 studies, identifying rapid progress but also remaining research gaps. One such gap is the limited use of domain knowledge on machine scheduling in current DRL-based solutions. This gap is addressed by proposing the incorporation of features inspired by the shifting bottleneck heuristic into the input of DRL agents. Although this particular method shows no significant benefit of the additional information, this pioneering approach is a positive impulse for future work in this overlooked research direction. A substantial contribution is further made regarding the training process. This thesis introduces a curriculum learning strategy for training DRL agents by drawing inspiration from latest AI research. The strategy involves a newly developed difficulty metric to sort training instances. The results suggest that starting the training with more difficult problem instances improves the agent's performance compared to conventional, unstructured training. The literature review further revealed a lack of emphasis on how trained agents can be used more effectively. This gap is addressed through the development of two innovative techniques. The first, called δ-sampling, manipulates the action output of a trained agent to promote better coverage of the solution space. The coverage is optimized considering the given number of solution candidates generatable in acceptable computational time, resulting in better schedules without additional training. A second technique applies decision transformers (DTs) that learn from interaction data between a trained base DRL agent and its environment. Due to their larger input context, DTs can incorporate information of how their previous actions led to the current intermediate solution, allowing them to outperform the trained base DRL agents.

Finally, this dissertation examines the transition from solving academic benchmarks that dominate current literature to real-world scheduling systems, analyzing additional challenges and proposing methodological approaches to overcome them. Special attention is given to promoting the reliability of future scheduling systems by deriving suitable definitions and corresponding metrics, as well as presenting promising ways to increase reliability in these terms.

By exploring different areas of improvement, proposing novel effective techniques, and outlining necessary future research, this work makes a decisive contribution to improving DRL-based machine scheduling systems and paves the way for their application in complex real-world production conditions.

Kurzfassung

Die industrielle Produktion wird durch komplexere Lieferketten, kürzere Entwicklungszyklen und höhere Produktvarianz vor wachsende Herausforderungen gestellt. Gleichzeitig sind mit der Verfügbarkeit von Produktionsdaten und Fortschritten in künstlicher Intelligenz (KI) neue Optimierungsmöglichkeiten gegeben. Motiviert durch die Relevanz der Herausforderungen und KI als neuem, vielversprechenden Lösungsansatz widmet sich diese Dissertationsschrift dem Einsatz von Deep Reinforcement Learning (DRL) für die Maschinenbelegungsplanung (MBP). Das erklärte Ziel sind DRL-basierte Planungssysteme, die bessere Pläne in kürzerer Rechenzeit generieren.

Die Arbeit beginnt mit einer umfassenden Literaturrecherche von 143 Publikationen, die zwar rapide Fortschritte aber auch persistierende Forschungslücken aufzeigt. So vernachlässigen bestehende Verfahren beispielsweise die Verwendung vorliegenden Domänenwissens aus der MBP. Daher wird in dieser Arbeit vorgeschlagen, etablierte Bottleneck-Informationen als Entscheidungsgrundlage für DRL-Agenten zu nutzen. Obwohl die konkrete vorgestellte Methode keine signifikante Verbesserung der Ergebnisse aufweist, dient sie als Pionierstudie dennoch als positiver Impuls für weitere Forschungsarbeiten in diesem vernachlässigten Bereich. Ein substanzieller Beitrag dieser Dissertation besteht in der Vorstellung eines innovativen Trainingsansatzes, inspiriert vom Curriculum Learning in der KI-Forschung. Der Ansatz beinhaltet ein neu entwickeltes Maß für die Schwierigkeit von Probleminstanzen, nach welchem Trainingsdaten sortiert werden. Es wird gezeigt, dass die Platzierung besonders schwieriger Probleminstanzen zu Beginn des Trainings zu besseren Ergebnissen führt als konventionelles, unstrukturiertes Training. Die Literaturrecherche zeigt zudem, dass bislang wenig Untersuchungen zur effektiven Nutzung bereits trainierter DRL-Agenten existieren. Daher werden hierzu in dieser Arbeit zwei neue Techniken vorgestellt. Die erste, das sogenannte δ-Sampling, manipuliert gezielt die prädizierten Aktionsvektoren der Agenten. Dadurch wird der Lösungsraum von den innerhalb einer akzeptablen Rechenzeit generierbaren Lösungen besser abdeckt. Somit führt δ-Sampling ohne weiteres Training zu besseren Plänen in der Nutzungsphase. Die zweite Technik nutzt sogenannte Decision Transformer (DT), die aus Interaktionsdaten zwischen einem trainierten Basis-DRL-Agenten und dessen Umgebung lernen. Der DT berechnet Entscheidungen im größeren Kontext seiner bisherigen Aktionssequenz mit einer Probleminstanz. Somit können Zusammenhänge in der Probleminstanz ausgenutzt werden, was zu besseren Plänen führt als die des ursprünglichen Basis-DRL-Agenten.

Zuletzt analysiert diese Arbeit zusätzliche Herausforderungen, die auf dem Weg von akademischen Untersuchungen hin zur praktischen Nutzung zur Lösung realweltlicher Probleme auftreten werden. Zudem werden methodische Herangehensweisen zu deren Bewältigung vorgestellt. Ein Schwerpunkt liegt dabei auf der Messbarkeit und der Stärkung der Zuverlässigkeit zukünftiger Planungssysteme.

Durch die Schaffung neuer, effektiver Methoden zur Erstellung besserer Pläne in kürzerer Rechenzeit sowie durch die Skizzierung notwendiger zukünftiger Forschungsarbeiten leistet diese Dissertation einen entscheidenden Beitrag zur Verbesserung DRL-basierter MBP-Systeme und ebnet den Weg für deren zukünftige praktische Anwendung unter komplexen realen Produktionsbedingungen.

Contents

1 Introduction 1
1.1 Motivation 1
1.2 Research Scope 4
1.3 Thesis Structure 5

2 Preliminaries 9
2.1 The Job Shop Scheduling Problem (JSSP) 10
2.1.1 Production Scheduling 10
2.1.2 Formalization and Characteristics of the JSSP 13
2.1.3 Common Methods for Solving the JSSP 19
2.2 Core Concepts of Deep Reinforcement Learning (DRL) 23
2.3 DRL for Job Shop Scheduling (JSS) 28
2.3.1 Necessary Design Decisions for DRL-based JSS 29
2.3.2 The Challenges and Vision of DRL-based JSS 30

3 Review of the State of the Art 33
3.1 Literature Corpus Generation 35
3.2 Results of the Review 38
3.2.1 Coverage of Job Shops and Related Scheduling Problems in the DRL Literature 39
3.2.2 Performance of State-of-the-Art Methods 47
3.2.3 Common Designs and Design Trends 51
3.3 Reflection on the State of the Art 65

4 **Domain Knowledge Integration in the Observation Space** 69
4.1 Related Work: Bottlenecks in JSSPs 72
4.1.1 Bottleneck Metrics and the Shifting Bottleneck Heuristic 72
4.1.2 Bottlenecks in DRL-based JSS 74
4.2 Experiment Design 75
4.2.1 Bottleneck Feature 75
4.2.2 Base Models and Bottleneck Feature Integration 76
4.3 Results 81
4.4 Summary and Discussion 85

5 **Effective Learning Curricula for the JSSP** 89
5.1 Related Work: Curriculum Learning for JSS 92
5.2 Experiment Design 96
5.3 Results 101
5.4 Summary and Discussion 105

6 **Effective Inference Strategies with Trained Agents for the JSSP** ... 109
6.1 Action Sampling in JSS 110
6.1.1 Intuition 111
6.1.2 Related Work: Action Sampling in JSS 113
6.1.3 Experiment Design 116
6.1.4 Results 119
6.1.5 Summary and Discussion 126
6.2 Decision Transformers for the JSSP 128
6.2.1 Related Work: Decision Transformers for the JSSP 129
6.2.2 Experiment Design 132
6.2.3 Results 142
6.2.4 Summary and Discussion 149

7 **From the JSSP to Real-World Applications** 151
7.1 Additional Challenges and Promising Remedies 153
7.2 Reliability of DRL-based Machine Scheduling 158
7.2.1 Definitions of Reliability in Machine Scheduling and DRL 159
7.2.2 Methods to Increase the Reliability of DRL-based Machine Scheduling 168
7.3 Summary and Discussion 171

8 **Critical Discussion and Outlook** 175

9 Summary 183

Bibliography 187

Abbreviations

A2C	Advantage Actor-Critic
A3C	Asynchronous Advantage Actor-Critic
ABZ	Adams, Balas, and Zawack benchmark dataset
ACL	Adversarial Curriculum Learning
AI	Artificial Intelligence
AS	Active Search
ASCL	Adaptive Staircase Curriculum Learning
ATTN	Attention-based Artificial Neural Network Architecture
AVaR	Average Value at Risk
CEI	Critical End Improvement Operator
CET	Critical End Time Operator
CL	Curriculum Learning
CNN	Convolutional Neural Network
CP	Constraint Programming
CPU	Central Processing Unit
CT	Critical Time Operator
CVaR	Conditional Value at Risk
DCT	Da Col and Teppan benchmark dataset
DDPG	Deep Deterministic Policy Gradient
DL	Deep Learning
DMU	Demirkol, Mehta, and Uzsoy benchmark dataset
DQN	Deep Q-Network
DRL	Deep Reinforcement Learning
DT	Decision Transformer

DT_{100}	Decision Transformer trained on data of 100 search iterations
DT_{200}	Decision Transformer trained on data of 200 search iterations
DTS	Difficulty To Solve
ECET	Extended Critical End Time Operator
ES	Evolution Strategies
FCN	Fully Connected Network
FT	Fischer and Thompson benchmark dataset
GA	Genetic Algorithm
GNN	Graph Neural Network
GPU	Graphics Processing Unit
ICL	Incremental Curriculum Learning
IT	Information Technology
JSP	Job Shop Problem
JSS	Job Shop Scheduling
JSSP	Job Shop Scheduling Problem
L2g	Learning To Generalize Method
LA	Lawrence benchmark dataset
MCTS	Monte Carlo Tree Search
MDP	Markov Decision Process
MILP	Mixed Integer Linear Programming
ML	Machine Learning
MTR	Most Tasks Remaining
NCO	Neural Combinatorial Optimization
NLS	Neural Local Search
NLS_{A}	NLS with Acceptance Decision Action Set
NLS_{AN}	NLS with Acceptance Decision and Neighborhood Action Set
NLS_{ANP}	NLS with Acceptance Decision, Neighborhood and Perturbation Action Set
NN	(Artificial) Neural Network
NP-hard	Non-deterministic Polynomial-time hard
OEE	Overall Equipment Effectiveness
OR	Operations Research
ORB	Applegate and Cook benchmark dataset
OV	Objective Value
PC	Production Control
PDR	Priority Dispatching Rule
POMO	Policy Optimization with Multiple Optima
PPC	Production Planning and Control
PPO	Proximal Policy Optimization

PR	Priority Rule
PS	Production Scheduling
RASCL	Reinforced Adaptive Staircase Curriculum Learning (formerly: ACL)
RL	Reinforcement Learning
RLHF	Reinforcement Learning from Human Feedback
RNN	Recurrent Neural Network
RQ	Research Question
SA	Simulated Annealing
SAC	Soft Actor-Critic
SBH	Shifting Bottleneck Heuristic
Sim2real gap	Simulation to Reality Gap
SPT	Shortest Processing Time First
SWV	Storer, Wu and Vaccari benchmark dataset
TA	Taillard benchmark dataset
TRPO	Trust Region Policy Optimization
TS	Tabu Search
TT	Trajectory Transformer
UCT	Confidence Bounds for Trees
VaR	Value at Risk
YN	Yamada and Nakano benchmark dataset

List of Figures

Fig. 1.1	Structure of chapters in this thesis	6
Fig. 2.1	Relationship between Production Planning and Control, Manufacturing Control, and Scheduling	10
Fig. 2.2	Structure and Task Reference View of the Aachen PPC Model, adapted from [12]	11
Fig. 2.3	Exemplary 3×3 problem instance (left) and a solution to it as Gantt chart (right)	14
Fig. 2.4	Disjunctive graph representation of a JSSP (left) and one example solution (right)	16
Fig. 2.5	Interaction pattern of a (D)RL agent and its environment	23
Fig. 2.6	Taxonomy of DRL algorithms (adapted from [48])	25
Fig. 2.7	Overview of the necessary design decisions of DRL solutions	29
Fig. 2.8	Vision of end-to-end DRL based machine scheduling	32
Fig. 3.1	Multi-step process for the literature corpus generation	37
Fig. 3.2	Number of publications on DRL-based machine scheduling between 2018 and 2025	40
Fig. 3.3	Scientific outlets of all publications in the literature review	42
Fig. 3.4	Ratios of the numbers of publications per outlet focus sorted by scheduling problem class	43
Fig. 3.5	Scientific outlets of publications on the JSSP in the literature review	44

Fig. 3.6 Counts of major contribution categories in articles on JSPs ... 46
Fig. 3.7 Ratio of the usage of benchmark data in JSSP studies (left) and counts of usages of JSSP benchmark datasets (right) ... 48
Fig. 3.8 Achieved average optimality gaps on the TA benchmark dataset over their publication date ... 50
Fig. 3.9 Counts of single- and multi-agent systems introduced over the years ... 52
Fig. 3.10 Counts of action logics used in the JSSP literature ... 54
Fig. 3.11 Iteration steps of the action-per-timestep and Tetris-like action logics ... 55
Fig. 3.12 Relative share (left) and counts over time (right) of learned and enforced action validity designs ... 58
Fig. 3.13 Relative share (left) and counts over time (right) of observation spaces consisting of engineered or raw features ... 60
Fig. 3.14 Counts of artificial neural network elements featured in studies over the years ... 62
Fig. 3.15 Counts of dense and sparse reward designs over the years ... 63
Fig. 3.16 Counts of used DRL algorithms sorted by policy-gradient-based and value-based algorithms ... 64
Fig. 4.1 Common steps of scheduling strategies under bottleneck considerations ... 71
Fig. 4.2 Bottleneck feature integration and data flow in the self-designed base model ... 78
Fig. 4.3 Bottleneck feature integration and data flow in the l2g base model ... 80
Fig. 4.4 Learning curves of the self-designed and l2g base models ... 82
Fig. 5.1 Intuition that easy tasks have an easy-to-find minimum in CL (adapted from [230]) ... 91
Fig. 5.2 Results of different CL strategies achieved by Iklassov et al. (adapted from [75]) ... 95
Fig. 5.3 Histogram of the DTS values of 40,000 training instances ... 99
Fig. 5.4 The four building blocks derived from sorting all 40,000 training instances by their DTS measure ... 100
Fig. 5.5 Curricula depicting the sequences of training instances sorted by DTS measure ... 100

Fig. 5.6 Evaluation scores across training according to different curricula 102
Fig. 5.7 Zoom-Ins on the second halves of agent training evaluation scores ordered by the common respective first curricular blocks 104
Fig. 5.8 Bar plots showing the relative and absolute immediate impact of introducing the second curricular blocks in training 105
Fig. 6.1 Illustration of the shift in distribution from exploration to exploitation for an exemplary action space size of three 112
Fig. 6.2 Hypothesis on the effect of sampling exploratively and exploitatively with different sample sizes 113
Fig. 6.3 Three iterations of the proposed search algorithm for δ^* 117
Fig. 6.4 Minimal makespans over sampling sizes for different sampling strategies 120
Fig. 6.5 Examples of results of the δ-value search algorithm with found minima 122
Fig. 6.6 Comparison of sampling methods over sample sizes 126
Fig. 6.7 Experiment workflow of the Decision Transformer study 133
Fig. 6.8 Data flow through the transformer model of the Decision Transformer (inspired by [237]) 134
Fig. 6.9 NN architecture of the NLS approach (adapted from [42]) 137
Fig. 6.10 Combination of the NLS model and the transformer implementation for the DT (parts adapted from [43]) 141
Fig. 6.11 Comparison of achieved average makespans by NLS and DT models with respect to the number of iterations (a) and the total search time in seconds (b) 146
Fig. 6.12 Action frequencies of NLS and DT models 147
Fig. 6.13 Mean makespans of the DT on 100 15×15 JSSP instances over varying return-to-go factors 148
Fig. 7.1 Interactions in a real-world DRL-based scheduling system 153
Fig. 7.2 Example for the reliability of two schedules under uncertainty 161

List of Tables

Table 2.1 Overview of common benchmark datasets for the JSSP 18
Table 3.1 Publications analyzed in the literature review sorted by problem category 38
Table 4.1 Hyperparameters of the self-designed base model 79
Table 4.2 Summary of average test results of the trained models 83
Table 6.1 Best found δ-values for different base models, problem sizes and sample sizes 121
Table 6.2 Average m^*-values on 100 generated test instances per problem size 124
Table 6.3 Average m^*-values on TA benchmark instances 125
Table 6.4 Action sets of teacher models. Those used in this study are marked with an “X“. 138
Table 6.5 Hyperparameters and design choices of the transformer model in the DT [43] 140
Table 6.6 Results (makespans and optimality gaps) on the Taillard benchmark instances 143
Table 6.7 Results (makespans) on 100 randomly generated test instances 145
Table 7.1 Overview of proposed reliability metrics 167
Table 7.2 Publications considering different notions of reliability 168

Introduction 1

1.1 Motivation

"Of Chess it has been said that life is not long enough for it, but that is the fault of life, not Chess."

This quotation is attributed to William Ewart Napier, a British-American chess master in the early 20th century. It implies that one can never perfect the game of chess. It is this sentiment of human intellect versus highly complex problems that has attracted people towards the game for centuries, and it is why grandmasters are by many considered to be geniuses. Over a hundred years later, in 2017, this sensation explains the sensational character and large media attention, when AlphaZero, a computer algorithm devised by the company OpenAI, was able to learn chess and other board games from scratch and beat human grandmasters as well as expert systems [1].

The triumph of AlphaZero in chess only four years after games in the Atari Computer Game suite had been solved with the same underlying machine learning paradigm, namely Deep Reinforcement Learning (DRL), significantly heightened the interest in applying DRL to complex industrial use cases. In theory, DRL is applicable to almost any sequential decision-making process. One such sequential decision-making process that occurs frequently at production sites is machine scheduling. Simply said, this omnipresent challenge deals with the allocation of resources to production tasks over time with the objective of optimizing production key performance indicators.

C. Waubert de Puiseau, *Advancing Training and Inference Methods for Reinforcement Learning-Based Job Shop Scheduling*,
https://doi.org/10.1007/978-3-658-51305-4_1

At their core, a large proportion of machine scheduling problems can be classified as combinatorial optimization problems, meaning that solutions are found within a finite set of discrete feasible solutions. Many variants of machine scheduling problems, including the job shop scheduling problem (JSSP), a special case of machine scheduling problems, are proven to be non-deterministic polynomial-time hard (NP-hard) for cases involving more than two machines [2]. As a result, the solution space grows exponentially with the number of resources and tasks that must be considered, and no known algorithm can enumerate and evaluate a sufficient number of feasible solutions within a polynomial time bound in order to ensure the arrival at an optimal solution. While it is therefore in practice inherently challenging to solve machine scheduling problems optimally, conventional methods for industrial scheduling perform more than 20 % worse than optimal [3], leading to suboptimal machine utilization and productivity. Given the substantial benefits that can be gained from optimizing schedules, even by a mere few percent at industrial scales, the search for better and more practical scheduling methods is an ongoing endeavor within both academic and industrial contexts.

This search has brought forth a plethora of dispatching rules, which revolve around determining priorities of jobs and selecting the highest priority for every decision that is made [3]. These priority rules are generally fast as well as straight forward to understand and implement, yet they do not tend to produce satisfying schedules on average. In a second research stream, numerous metaheuristics have been and are still developed, aiming to strike a better balance between computational time and optimality [4]. A third category of solution algorithms comprises numerical solvers, which perform a structured search over the whole solution space by means of sophisticated branch and bound techniques [5]. On the upside, this enables solvers to find optimal solutions with optimality guarantees. On the downside, exact solvers become intractable as problem sizes increase for NP-hard problems.

In recent years, following the aforementioned breakthroughs in board and video games, deep reinforcement learning (DRL) has become increasingly popular as a data-driven self-learning alternative for machine scheduling problems [6].

In fact, a whole new research area, termed *neural combinatorial optimization* (NCO), has emerged that addresses machine scheduling and related NP-hard combinatorial optimization problems. The growing interest observable in academia is driven by three primary characteristics of DRL-based approaches that make it advantageous over existing approaches in this problem class:

1. It offers potential for a favorable trade-off between the attainment of better solutions and longer required computation times. This is because computation time may be spent offline during training, before the agent is actually used.
2. Production facilities are dynamic and multifaceted environments, which, despite all efforts, are subject to fluctuating conditions such as machine availability or human performance. Such stochastic nuances are difficult to quantify and to model explicitly, but could be learned implicitly from experience data.
3. The algorithm is self-adaptive, such that changes in the production facility, once integrated into the learning environment, are automatically addressed through fine-tuned training, thereby reducing the required manual effort to adjust prioritization rules.

The vision of self-adapting DRL-based scheduling systems that learn directly from and are applied to real-world use cases is actively pursued by a fast-growing research community. First attempts date back to 2018 [7]. Since DRL-based machine scheduling lies at the intersection of many different established research domains such as operations research (OR), production research, informatics, computer science and artificial intelligence (AI), the problem has been approached from different perspectives. The two major perspectives represented in the literature may be categorized into basic research on academic combinatorial optimization problems and applied research on single specific use cases. In general, those with an inclination towards the former perspective are faster to adopt new trends and achievements from the field of AI research, such as novel learning paradigms or innovative artificial neural network (NN) architectures. However, their focus is primarily on abstract, academic problems. In contrast, research from the latter perspective incorporates more constraints and dynamic factors into the learning problems, thereby identifying new challenges and opportunities for real-world applications. Nevertheless, these advanced problems are typically tackled with less sophisticated DRL-methods that might be regarded as already outdated in the AI community.

Both perspectives present great opportunities for advancing the field of DRL-based machine scheduling. This thesis aims to contribute to both research streams and bring them closer together. On the one hand, opportunities of integrating established methods from the AI and OR domains into DRL-based machine scheduling algorithms are identified and evaluated. On the other hand, practical notions of the usage of trained DRL agents are considered and new methods derived. Finally, the path from achieving very competitive results on academic problems to tackling real-world problems is analyzed and suggestions for the way forward are outlined.

1.2 Research Scope

The overarching objective of this thesis is to increase the effectiveness of DRL-based machine scheduling and thereby making it a more competitive solution in terms of obtaining shorter makespans in shorter computation times. To this end, this thesis poses three research questions. They address three different types of design decisions that arise during the development of DRL-based machine methods. The JSSP is used as representative object of study for machine scheduling problems.

The first examined design decision is that of the observation space. The information basis for DRL agents typically comprises simplistic descriptive and statistical features of the jobs, resources and the partially solved schedule. This is further detailed in Chap. 3. Regarding the observation space, the objective pursued in this thesis is to utilize existing algorithms and expert knowledge from the OR domain by deriving informative features at each intermediate state of the decision-making process from which the agent can learn and which it can use to make informed decisions. In particular, the focus lies on the common notion of *bottlenecks*. An advantageous outcome would be faster convergence speed during training or shorter makespans during inference. The research question (RQ) that is addressed is:

RQ1 How can domain knowledge be advantageously integrated into the observation space?

For the JSSP, as a problem of combinatorial state space size, it is non-trivial to cover sufficient relevant sample instances during training, such that the agent may derive a generalizing solution strategy. It is established that some problem instances are solved near-optimally with simple dispatching rules, while others are not [3]. It is also intuitive that dispatching strategies should vary in accordance with the problem size of the JSSP instance to be solved. Similar considerations have been previously raised for other application domains of DRL as well. These are nowadays subsumed under the umbrella of *curriculum learning* (CL) [8]. CL aims to vary the difficulty of the learning task over the course of training for the purpose of faster convergence and better final policies of the agents, i.e., greater training efficiency and effectiveness. Since CL for the JSSP has received little attention in the literature, the following research question is addressed:

RQ2 How should training strategies be designed for efficient and effective DRL agent training?

Once an agent has been trained effectively, it can be used to evaluate any given state in the planning process. To generate a plan, one couples the agent with a simulation of a machine scheduling environment, initializes the current planning state and follows the decisions of the agent directly. However, with greater computational time available, one may simulate multiple scenarios and find a better plan amongst all generated solutions. In essence, in a subsequent step to agent training, the following question arises:

RQ3 How should a trained DRL agent be used, given a fixed computational budget during inference, to decrease the expected value of the minimal makespan of the generated solution?

The aforementioned questions primarily address methods for optimizing an objective value for JSSPs in simulation for academic rather than practical use. However, on the path to the fulfillment of the vision of RL-based production scheduling in the real world, another crucial aspect is the reliability of the system, such that uncertainties are dealt with, and users may trust a scheduling agent enough to follow its recommendations. Therefore, to establish a foundation for future research and development, the following fourth open research question is posed:

RQ4 Which additional challenges arise when transitioning from the abstract JSSP to real-world production planning scenarios and how may they be addressed?

1.3 Thesis Structure

The thesis is structured in a linear fashion, comprising nine chapters, as illustrated in Fig. 1.1. After this introduction, the prerequisites for this work are presented in Chap. 2. The prerequisites provide an overview of production scheduling problems and the characteristics of the object of study, namely the JSSP, including the variety of methods to solve the JSSP. Subsequently, the fundamental concepts of DRL are summarized, and possible design decisions that must be made when applying it to JSSPs are categorized. These categories provide the context into which the research questions are embedded.

Fig. 1.1 Structure of chapters in this thesis

Chapter 3 examines the rapidly evolving state-of-the-art of DRL-based job shop scheduling (JSS). To this end, the chapter analyzes 143 related publications of the last six years. It compares the design decisions made by fellow researchers, and highlights promising trends in this domain. Given that the JSSP presents only one specific class of problems within many related scheduling problems, it is worth considering the literature of DRL applications to related problems, too. Consequently, the range of categories of scheduling and control problems in the literature corpus is first expanded beyond the strict JSSP, and then the design decisions applicable to the JSSP are extracted. Complementary to the overall literature review, more specific analyses of the state of the art and related work of the studies presented in this thesis are featured at the beginning of each of the respective Chaps. 4–6.

Chapter 4 addresses RQ1 by reviewing different definitions and metrics that have been proposed for bottlenecks in job shops, including the definition used in the so-called shifting bottleneck heuristic (SBH) [9]. A promising variant is derived and incorporated into the observation space of two distinct DRL-based JSS approaches. Then, experiments are conducted to ascertain whether the additional feature is beneficial for the efficiency and effectiveness of model training. In Chap. 5, different learning curricula are demonstrated, in which training instances are sorted based on the difficulty to solve them well with a priority dispatching rule. The curricula are experimentally evaluated, and the learning behavior is thoroughly analyzed to gain insights into the effective mechanisms of the proposed curricula. As such, Chap. 5 sheds light on RQ2. Chapter 6 explores RQ3, namely, how trained agents are optimally used during inference. To answer this question, a budget-aware trajectory sampling strategy for the application to the JSSP is developed and analyzed.

Chapter 7 builds upon the preceding methods and addresses the open RQ4 dealing with additional challenges, which arise when DRL-based JSS is transferred into real-world production. A particular focus is put on reliability considerations. As a foundation for future work, the differences and overlap of interpretations of reliability are examined in the two domains of OR and DRL, metrics are derived and promising research directions to increase the reliability of DRL-based JSS systems are outlined.

Chapter 8 presents a critical discussion of the overall contributions described in the different sections of this work. It identifies open research gaps, and proposes potential avenues for future research as well as next steps that are necessary from the current state of research to make the vision of DRL-based production scheduling a reality. Finally, the thesis is summarized in Chap. 9.

Preliminaries

2

This chapter serves as a formal introduction to the JSSP, the central object of study in this thesis, and to DRL, the central investigated solution paradigm. It also introduces the implications of applying DRL to the JSSP. The structure of this chapter is as follows:

The first subchapter, 2.1, provides the real-world context of production scheduling and control problems, emphasizing their real-world significance, occurrence, and objectives, and summarizing the challenges that arise when solving them. It also introduces a general formal categorization of production scheduling problems, before detailing different ways to conceptualize and formalize the characteristics of the JSSP in particular. The first subchapter concludes with a brief overview of common methods for solving the JSSP as well as their respective limitations. This serves to deepen the understanding of the problem, and to pinpoint DRL in the landscape of JSSP solutions.

The second subchapter, 2.2, summarizes the core concepts and theoretical foundations of DRL. Together, the first two subchapters provide background information for subchapter 2.3, which explains the alignment of the object of study, the JSSP, and the applied solution methodology, DRL. The explanation includes the necessary design decisions for DRL-based JSS to facilitate the understanding of how this alignment is realized in Sect. 2.3.1. Additionally, Sect. 2.3.2 provides initial insights into the challenges of addressing JSS with DRL but also elaborates on the vision of DRL-based JSS to highlight the promising long-term potential that this combination holds.

C. Waubert de Puiseau, *Advancing Training and Inference Methods for Reinforcement Learning-Based Job Shop Scheduling*,
https://doi.org/10.1007/978-3-658-51305-4_2

2.1 The Job Shop Scheduling Problem (JSSP)

2.1.1 Production Scheduling

Production Scheduling (PS) is a subfield of production control (PC, also referred to as manufacturing control), which is in turn a subfield of production planning and control (PPC), although the three terms are sometimes used interchangeably [10]. Their relationship is illustrated in Fig. 2.1.

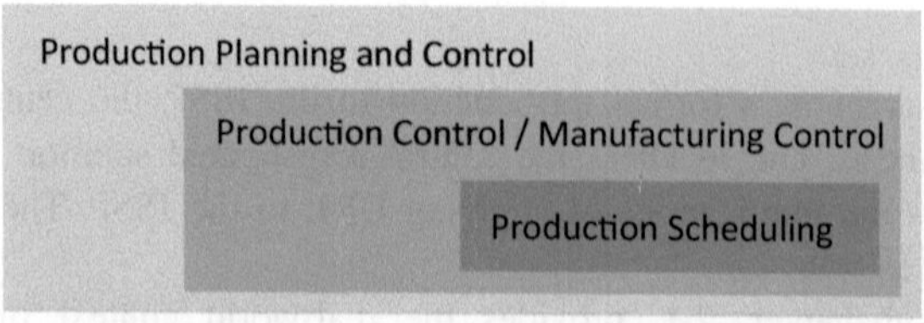

Fig. 2.1 Relationship between Production Planning and Control, Manufacturing Control, and Scheduling

Within this framework, PS deals with the planning of resource utilization. A broader definition of scheduling, which is not limited to the production setting, is given in the well-known foundational work "Scheduling. Theory, Algorithms, and Systems" by Michael L. Pinedo [11]: "Scheduling is a decision-making process that is used on a regular basis in many manufacturing and services industries. It deals with the allocation of resources to tasks over given time periods and its goal is to optimize one or more objectives" [11, p. 1].

In the 607-page book, this definition is followed by several examples that illustrate the wide range of scheduling problems in terms of their application scenario, objective, and urgency. In these examples, PS as an application domain forms the largest subcategory. In the PS subcategory tasks (also referred to as operations) are production steps applied to industrial products, and resources are often production machines. When multiple operations are related, for example because they are performed on the same physical part, they form a job. Production control extends pure PS by the inclusion of storage decisions [10]. PPC, in contrast to PS and PC, is concerned with production decisions on a higher strategic level, such as dealing with the timely adjustment of the production rate to market demands, external parts of the supply chain, or production requirements planning to ensure that necessary production resources are generally available.

The reason for the interchangeable use of the terms PS, Production Control (PC) and PPC may be due to the fact that many individual tasks within PPC have an indirect or even direct impact on others, as illustrated in the Aachen PPC Model [12, p. 16] shown in Fig. 2.2. The Aachen PPC model focuses on PS as a separate task from PPC and PC. However, since they are sometimes tightly linked, certain elements of PPC will appear in some of the applied studies mentioned in the literature review in Chap. 3.

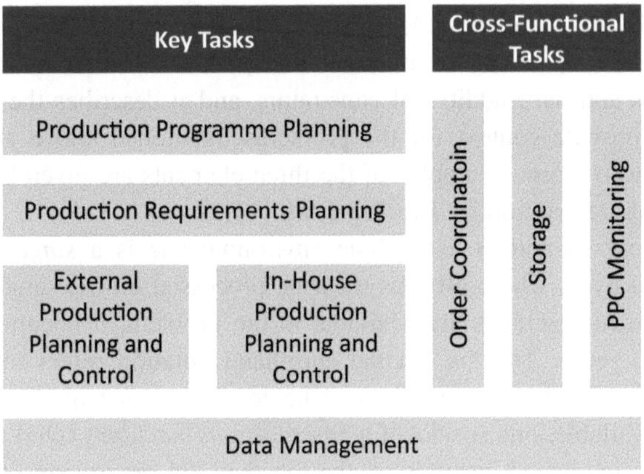

Fig. 2.2 Structure and Task Reference View of the Aachen PPC Model, adapted from [12]

From a practical perspective, PS can be performed online or offline. Offline scheduling is applied in problem scenarios, where all problem data is available in advance, such that the entire schedule can be created with perfect information [11]. In a real-world setting, this would equate to, for example, creating a one-week production schedule on a Monday morning, given the current order status with a backlog that requires a total of three weeks to complete. In online scheduling, information about the number of tasks and their characteristics only becomes available over the course of creating a schedule [11]. In practice, in the case of online scheduling, new orders arrive several times per day and take only minutes to produce, making a strict long-term plan infeasible. Therefore, in such a case, decisions are rather made on an event-by-event basis, for example, whenever any machine becomes idle. In theory, it is possible to separate problem scenarios into online and offline scheduling. However, in practice, circumstances are likely to change, even in a mostly static environment. It is in the nature of production that some new tasks will arrive over time, or delays will occur due to resource failures while the original production plan is being followed. Consequently, scheduling and frequent rescheduling to adapt to new circumstances is a common practice that uses offline scheduling as an initial basis but also integrates more reactive online scheduling to form a complete method. [11]

A common formal notation for machine scheduling problems is the three-field ($\alpha|\beta|\gamma$) notation introduced by Graham et al. [13], which has also been adapted in the work of Pinedo [11]. This notation has recently been proposed as a suitable formalization in the DRL-based machine scheduling literature by Rinciog et al. [14] as well. In this notation, α describes the machine environment with machine capabilities with respect to jobs and operations, β describes the characteristics of the set of jobs and their additional constraints, and γ describes the optimization criterion. To provide context for the problems mentioned above and discussed later in this thesis, some examples of the three elements are given below, largely inspired by the description by Rinciog et al. [14].

The simplest version of a machine environment α is a *single-machine* (1) setup, denoted by 1, where all jobs must be processed on the same resource. If multiple identical machines are available in the problem, it becomes a *parallel machine* (*Pm*) setup. The "*m*" in the shorthand notation refers to the number of machines involved in the problem. If heterogeneous machines with different abilities are available, one speaks of a *heterogenous machines* (*Qm*) problem, and if their processing speed depends on the job they are processing, it is called an *unrelated machines* (*Rm*) problem. Obviously, each problem is a subvariant of the previous one, and a problem could also be a hybrid of several. Whenever setups contain multiple machines and multiple jobs consisting of multiple operations, these setups are referred to as *shop-scheduling* setups. If the operations within a job have no precedence constraints, i.e., they do not need to be completed in a fixed order, but each operation requires a particular machine, the problem is called an *open shop* (*Om*) problem. If the order is fixed but identical in all jobs, the setup is called a *flow shop* (*Fm*). If the order is different in different jobs, the setup is called a *job shop* (*Jm*). The term "flexible" is added if an operation can be processed on more than one machine, which can lead to a *flexible flow shop* (*FFc*) or *flexible job shop* (*FJc*) problem. [14]

The additional constraints β comprise any further characteristics that affect the scheduling of jobs to machines. Frequently encountered examples are *limited buffer sizes* in front of machines (b_k), required *setup times* of jobs on machines where the setup times depend on the sequence of both the preceding operation on the machine and the next operation (s_{seq}), limited transport units ($trans_t$), or *batch processing* ($batch_b$). For some additional constraints there exist no dominant notations in the literature. Such constraints include but are not limited to stochastic machine breakdowns, stochastic job releases, any kind of stochastic parameters like processing times, or additional required resources. [14]

The third element of the three-field notation, γ, defines the objective of the scheduling problem. There are many different objectives to optimize for in reality, and many of them are correlated. A common example of a scheduling objective is the *completion time* (C_{max}), also known as the *makespan*, which measures the time between the start of the first operation and the completion of the last operation in the set of all jobs and operations. The completion time correlates strongly with the *machine utilization* (Utl) and *throughput times* (Tpt). Other, deadline or delivery time related objectives, include the *earliness* (E) and *tardiness* (T) of completed jobs, defined as the negative and positive difference between the completion dates of jobs to predefined deadline dates, respectively. Recently, the minimization of *material waste* (W) and *energy utilization* (Utl_e) has also received more attention [15, 16].

Although machine scheduling problems can be exactly defined through these three descriptive features, it can be difficult to draw distinct lines between certain setups, especially if multiple machine setups, constraints and objectives are combined and mixed within one problem.

2.1.2 Formalization and Characteristics of the JSSP

As the previous section highlighted, scheduling problems can be defined by their machine setup, additional constraints and objectives. In this thesis, the focus lies on job shop problems (JSPs). In terms of the three-field notation, this general class would be represented as $Jm|?|?$, where the set of entries for β and γ is undefined and can vary. Accordingly, JSPs encompass a wide variety of scheduling problems. Most studies in this thesis focus on a particular variant of JSPs that is one of the most studied in the literature [11] due to its easily defined constraints and objective. In this subsection, this variant is introduced, different representations of it are explained, and its characteristics are summarized. To distinguish the variant from the more general JSP problem class, it will be referred to as *the JSSP*, or *standard JSSP*, from here on.

The JSSP in Three-Field Notation, in Words and as Gantt chart

In the three-field notation, the JSSP is denoted by $Jm|p_{ij}|C_{max}$. This means that there exist m machines ($m \in M$) that work on j jobs ($j \in J$) each consisting of i

operations ($i \in I$) each. Since it is a job shop scheduling problem, in the JSSP, the order of the operations in each job is fixed through precedence constraints. Each operation has a fixed processing time p_{ij}. In most cases, the number of operations I and the number of machines M are equal, and every job visits each machine once. Thus, every operation in a job is processed on a distinct, fixed machine. The size of a JSSP instance can therefore be described through the number of its jobs J and its machines M by $J \times M$. A machine can only process one job at a time. Once started, an operation cannot be stopped but is processed without interruption, which is referred to as no-preemption constraint. One further assumes the existence of infinite buffers before and after machines and neglects transport times between machines. The objective is the minimization of the makespan C_{max}.

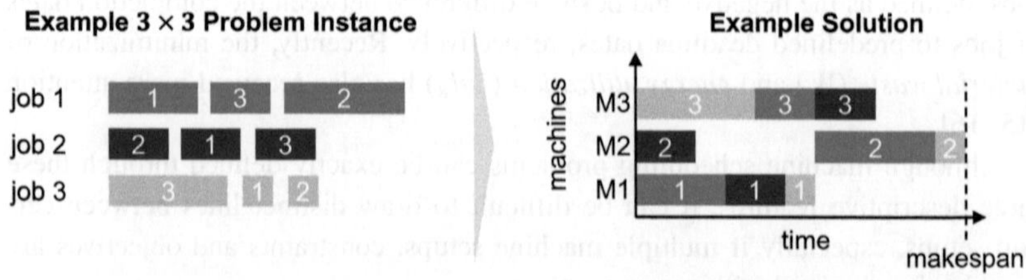

Fig. 2.3 Exemplary 3 × 3 problem instance (left) and a solution to it as Gantt chart (right)

Figure 2.3 shows an exemplary 3 × 3 JSSP instance (left), in which each job has a different color, and which consists of three ordered operations represented by individual blocks. The length of each operation is represented by the width of its block, and the machine on which it must be scheduled is represented by the number on the block. The right side of the figure shows one of many possible arrangements of the order of operations over time while adhering to all constraints as a Gantt chart. In this common visualization of scheduling solutions, machines are on the y-axis and the time is on the x-axis. The makespan is indicated by a dashed line.

The JSSP as Mixed Integer Linear Problem

A formal mathematical notation is that of a mixed integer linear optimization problem (MILP). One common representation as MILP is the disjunctive version used in [17] which is defined as follows:

$$Minimize C_{max} = \max_{j} C_j \tag{2.1}$$

$$s.t.: \quad x_{ij} \geq 0, \qquad \forall j \in J, i \in M \tag{2.2}$$

$$x_{\sigma_h^j, j} \geq x_{\sigma_{h-1}^j, j} + p_{\sigma_h^j, j}, \qquad \forall j \in J, h = 2, ..., m. \tag{2.3}$$

$$x_{ij} \geq x_{ik} + p_{ik} - V \cdot z_{ijk}, \qquad \forall j, k \in J, i \in M, j < k \tag{2.4}$$

$$x_{ik} \geq x_{ij} + p_{ij} - V \cdot (1 - z_{ijk}), \qquad \forall j, k \in J, i \in M, j < k \tag{2.5}$$

$$C_{max} \geq x_{\sigma_m^j, j} + p_{\sigma_m^j, j}, \qquad \forall j \in J \tag{2.6}$$

$$z_{ijk} \in \{0, 1\}, \qquad \forall j, k \in J, i \in M \tag{2.7}$$

Here, x_{ij} is the decision variable representing the start time of the operation of job j in all jobs J on machine i in all machines$M$. Further, p_{ij} is the corresponding processing time, all operations σ follow an order within a job (σ_1^j,σ_2^j, ..., σ_k^j), and z_{ijk} is the binary variable indicating that job j precedes job k on machine$i$. Constraint (2) ensures that each operation starts at a time greater than or equal to zero. Constraint (3) ensures that precedencies within jobs are enforced, and constraints (4), (5) and (7) ensure that operations are not scheduled onto the same machine at the same time if V is a large number. Constraint (6) defines the completion time as the time at which the last operation is completed. Note that this formulation using V is only one of several that are suitable for MILP solvers. A solution is uniquely defined through the start times x_{ij} of all operations.

The JSSP in Graph Notation

Another representation of the problem as well as partially solved problems is the formulation as disjunctive graph, as described by Błażewicz et al. [18]. An example of the same 3×3 JSSP instance described above and the same solution, both as disjunctive graphs, are shown in Fig. 2.4. Each node in the graph represents an operation of a job, with an additional dummy start node (s) and an additional dummy terminal node (t). The job nodes are colored according to the job to which they belong and appear in the same row. Operations of the same job are connected by directed edges shown as gray arrows, indicating the precedence constraints between nodes. The bold number in each operation node indicates the processing time whereas the indices below indicate the job number and the operation index within the job, respectively. An additional edge type is used to connect operations that must be processed by the same machine. These edges are shown as dashed lines and are originally undirected in an unsolved instance representation. To form a unique solution, the order of operations on each machine, i.e., the machine precedencies, must be determined. To do this, the originally undirected machine edges are transformed into directed edges, while unnecessary undirected edges are removed. This is shown on the right side. The graph representation also allows the definition of a partial solution, where only a subset of all machine edges is directed. Note that, once directed, the machine edges also represent a time ≥ 0 that lies between nodes that would correspond to white spaces in the Gantt chart in which the respective machine is idle.

The graph representation lends itself for matrix computations and can be easily used in conjunction with graph neural networks (GNNs), which aim to capture the instance characteristics through message passing between nodes along edges [19]. The makespan equates to the longest path in the graph, also referred to as the critical path. Note that his path is not directly visible but must be inferred from the processing times of nodes and times represented by directed edges.

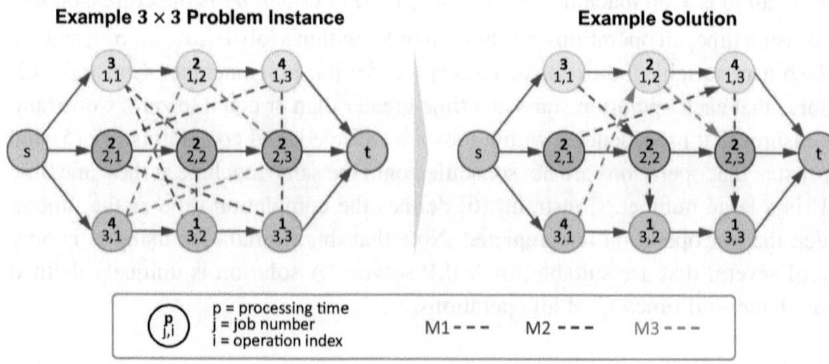

Fig. 2.4 Disjunctive graph representation of a JSSP (left) and one example solution (right)

Characteristics of the JSSP

The JSSP belongs to the class of combinatorial optimization problems, characterized by a search space of discrete elements and a resulting finite set of solutions [20]. The discrete nature of the elements in the search space is evident from looking at the Gantt chart visualization in Fig. 2.4, where one can intuitively see that swapping the order of two elements on a machine would result in a very different discrete solution. The JSSP in the form presented here has been proven to be NP-hard if it involves two machines or more [2]. This solution complexity class implies that no known algorithm can solve the JSSP in less than or equal to polynomial time with respect to the number of jobs and machines. With respect to the discrete search space, it is useful to calculate the theoretically possible number of solutions. Looking at either the Gantt chart in Fig. 2.4 or the graph representation in Fig. 2.5, it is evident that the number of possible solutions scales with the number of possible permutations of the order of operations on a single machine, but also with the number of machines. The number of permutations on a machine is the faculty of the number of jobs $|J|!$. Since one can combine any of these permutations on one machine with any permutation on another machine, the absolute number of combinations is exponentiated by the number of machines $|M|$ resulting in the absolute number of combinations expressed by $(|J|!)^{|M|}$. For a 3×3 problem, there are $3!^3 = 216$ combinations, and for a 6×6 problem already $6!^6 \approx 1.39 \times 10^{17}$ combinations. For this reason, enumerative algorithms that list all possible solutions quickly become infeasible for finding optimal solutions as problem sizes increase.

Due to its structure, two further properties of the JSSP are its problem symmetry and solution symmetry [21]. These concepts will be important for interpreting the results in Chap. 6. Problem symmetry refers to the property that the problem can be rewritten, but certain main characteristics would not change. For example, switching the job index between two jobs defines a new but equivalent problem. Solution symmetry refers to the fact that many different solutions, e.g. expressed through different starting times of operations, lead to the same objective value. This can also mean that there are multiple global optima. Symmetries can be used to make solution algorithms more efficient by leveraging them to eliminate parts of the solution space [21], but could have potential detrimental effects on learning-based algorithms that receive the same learning signals for very different solutions.

Benchmark Datasets

Several datasets of JSSP instances have been introduced and published over the last decades, some reporting proven optimal solutions and some not. Table 2.1 provides an overview of the most common datasets. These datasets are still used today to benchmark the performance of different solution algorithms in terms of how close they come to the optimum in what required computational time. The datasets differ regarding the problem sizes and the number of instances they contain, but also regarding the distribution from which processing times are sampled. For example, in the dataset by Taillard (TA) [22], which is used several times throughout this thesis, the processing times are sampled from a uniform distribution in the interval [0, 99]. The most recent addition to such benchmark datasets is the dataset introduced by Da Col and Teppan (DCT) in 2019 [23], which consists of much larger problem instances than previous datasets. The authors did this because they claim that smaller instances are very fast to solve with modern techniques, do not pose significant challenges, and because many alternative methods can achieve near-optimal solutions. Therefore, they do not leave enough room for demonstrating improvements of novel methods. The common methods for solving JSSP instances are summarized in the next subsection.

Table 2.1 Overview of common benchmark datasets for the JSSP

Common Abbrev.	Problem Sizes	Number of JSSP Instances	Year of Pub.	Authors and Publication
FT	6×6, 10×10,20×5	3	1963	Fischer and Thompson [24]
LA	10×5, 15×5, 20×5, 10×10, 15×10, 20×10, 30×10,15×15	40	1984	Lawrence [25]
ABZ	10×10, 20×15	5	1988	Adams, Balas and Zawack [9]
ORB	10×10	10	1991	Applegate and Cook [26]
SWV	20×10, 20×15,50×10	20	1992	Storer, Wu and Vaccari [27]
YN	20×20	4	1992	Yamada and Nakano [28]
TA	15×15, 20×15, 20×20, 30×15, 30×20, 50×15, 50×20,100×20	80	1993	Taillard [22]
DMU	20×15, 20×20, 50×15, 30×20, 40×15, 40×20, 50×15,50×20	80	1998	Demirkol, Mehta and Uzsoy [29]
DCT*	10×10, 10×100, 10×1000, 100×10, 100×100, 100×1000, 1000×10, 1000×100,1000×1000	90	2019	Da Col and Teppan [23]

* Own abbreviation

2.1.3 Common Methods for Solving the JSSP

This section outlines common methods for solving the JSSP that can be found in the literature. In this thesis, "solving" does not refer to solving instances optimally, i.e., finding the global minimum of the makespan, but it refers to finding a feasible schedule with the intention of achieving a small makespan. Generally, while finding feasible solutions for the JSSP is cheap, finding and identifying the provably optimal solution is very expensive. Solution methods can be characterized by how fast they can achieve a feasible makespan that is not far from optimal. The computational time and the degree of optimality are often seen as opposing criteria. From a practical point of view, when using solution methods in industry, two additional qualitative evaluation criteria for solution methods can be important. One of them is how easily the solution method can be implemented and adapted to slightly changing conditions on a shop floor. Methods that are difficult to implement and adapt to new problems require more time and expertise to use, which is disadvantageous. The other criterion is how easily the solution method can be interpreted. If a method allows practitioners to explain why a certain scheduling decision is made, the solution is more interpretable which can lead to more trust and acceptance by users [30].

Priority rules, which are applied in various ways to create solutions, are at one end of the spectrum of these four criteria. Rule-based priorities are cheap to compute, easy to implement and interpret, but on average they do not yield competitive makespans. Priority rules rank a set of jobs or machines according to some (often simple) calculations to allow a choice between them when a decision is needed. One example are Priority Dispatching Rules (PDRs) according to which one decides, which job is dispatched next into the job floor, or which job will be processed next by a machine that has just become idle. To create a solution with PDRs, the job floor is typically represented in a simulation in which priority rules are used to decide which jobs should start next on an idle machine. A complete solution is obtained when all operations of all jobs in the simulation have been scheduled. Two prominent examples of priority rules are SPT (Shortest Processing Time First) and MTR (Most Tasks Remaining). In SPT, jobs are sorted by the processing time of their next unfinished operation, and the one with the shortest value is prioritized. In MTR, as the name implies, jobs are similarly sorted by the number of remaining unfinished operations. A good summary of existing priority rules and their performance is given in the survey by Haupt et al. [3].

Although they are simplistic and far from optimal, priority rules are popular, and many variations are used in practice. The literature review conducted in

Chap. 3 identified 67 distinct priority rules, illustrating their popularity. When priority rules are used in the described combination with a simulation to generate schedules, they are sometimes also referred to as "construction heuristics" or "constructive heuristics" because solutions are generated, or constructed, operation by operation. One can think of construction heuristics as building a Gantt chart block by block or as adding directed arcs to a graph representation. This logic is what makes priority rules so fast, interpretable, and easy to implement. In fact, the same set of rules can be applied to a variety of scheduling problems outside of the JSSP. As mentioned earlier, the disadvantage of these rules is that they do not return schedules with small makespans compared to other solution methods [11].

The simulation-aided solution construction process can be visualized as a decision tree, where each decision between different jobs represents a node. A well-known alternative to making each decision using a priority rule is to apply Monte Carlo Tree Search (MCTS) [31]. MCTS is a stochastic search algorithm that simulates the outcomes of branches in the search tree (in case of the JSSP, this equates to the creation of complete schedules) and then uses this knowledge to evaluate decisions in the tree construction. Specifically, it follows an iterative four-step regime. It starts with the *selection*, where one node in the tree is selected as a candidate for further expansion. The selection is based on node values that are calculated according to a predefined strategy, such as following Upper Confidence Bounds for Trees (UCT). In the second step, the *expansion*, several child nodes are added to this selected node. From these new child nodes, one or multiple simulations are run until a complete solution is generated. Intermediate decisions are made either randomly or by following a predefined heuristic (such as a priority rule). This step is called *simulation* or *rollout*. These rollouts are used to backpropagate value estimates, which update node values in the partial tree that has been constructed thus far. Then, based on the new node values, UCT values can be updated, and a new iteration starts. The process continues until a complete decision tree, i.e., solution, is generated. [32]

At the other end of the above-mentioned criteria for solution methods are numerical solvers, such as MILP or Constraint Programming (CP). Both of these numerical solver paradigms comprise a large set of methods that are combined in software products such as Gurobi [33], IBM's CPLEX software, or Google OR-Tools [34], that are advancing over time. The main advantage of such solvers is that they can return provably optimal solutions, guaranteed by a structured search process. For small problem sizes, such as 6×6 or 10×10 JSSPs, optimal solutions can often be found within seconds or minutes on a consumer-grade laptop. However, for larger problem sizes, such as 15×15, there are cases where no optimal solution is found even within an hour of computation time. In other words, the NP-hardness of the JSSP takes its toll. It should be noted that most algorithms of this class keep a memory of the best feasible solution found, and if they are stopped before reaching an optimum, this best solution may already have a near-minimal or even globally minimal makespan. Yet, for problem instances of size 100×100 and above, numerical solvers fail to find feasible solutions even after running for several hours [23]. Thus, the required computation time is an advantage of these algorithms for small problems but becomes a disadvantage for large problems compared to construction heuristics. A clear disadvantage of numerical solvers, however, is the considerable implementation effort needed for adjustments if machine constellations and constraints change and become more complex to represent. Another disadvantage is the interpretability of individual scheduling decisions that are made, for which no rule-based explanations can be given.

A third class of solution methods can be summarized as metaheuristics [35, 36]. Metaheuristics are designed to guide the search in the solution space through heuristic or stochastic decisions, starting from one or multiple initial suboptimal schedules. Metaheuristics are flexible in terms of their required computational time because the search process can be stopped at any time. On the JSSP, the resulting makespans of metaheuristics fall somewhere between priority rule-based solutions and optimal solutions. Unlike numerical solvers, the search procedures that metaheuristics guide are not structured such that no optimality guarantees can be given. Metaheuristics often do not require a strict mathematical formalization of the problem but instead work as black-box algorithms that are indifferent to the internal structure of the problem. Instead, they work with intuitively understandable changes to existing solutions (schedules). This makes metaheuristics easier to implement than solvers, and also more intuitive for the user to some extent, because intermediate solutions can be examined. In contrast to priority rule-based solutions, metaheuristics are typically not tailored to a specific problem class, but are rather intentionally designed to be non-problem specific [35].

Metaheuristics can be further divided into trajectory-based metaheuristics and population-based metaheuristics [35]. Trajectory-based metaheuristics explore a single solution trajectory over time, starting from an initial solution and aiming to improve this solution by making local changes to it. There are different methods, each of which aims to traverse the search space as effectively and efficiently as possible through different tricks. A common method is Simulated Annealing (SA) [37], which is inspired by the annealing process in metallurgy, initially allowing larger changes to the solution and gradually decreasing changes for finer adjustments. This method aims at escaping local optima early in the process, but eventually optimizes efficiently towards a final local optimum [31]. Another well-known representative of trajectory-based metaheuristics is Tabu Search (TS), which uses a tabu list in memory to avoid revisiting already explored solutions, thereby eliminating ineffective iterations and mitigating oscillations in the search process [36].

Population-based metaheuristics maintain and evolve a set (population) of solutions, often drawing inspiration from natural or social systems to explore the solution space. The perhaps best-known group of algorithms in this class are called Genetic Algorithms (GAs), which are loosely inspired by biological evolution. They feature selection phases, gene crossovers and mutations that are responsible for optimizing a population of solutions according to a fitness function [38]. Another example of a population-based metaheuristic is Particle Swarm Optimization. It mimics social behavior in a swarm of solutions, in which neighboring solutions influence each other. This leads to a gradual optimization through the spread of information about local and global optima [39]. Many other variants of both trajectory-based and population-based metaheuristics exist. In addition, hybrid versions between different metaheuristics, numerical solvers, and construction heuristics can also be found in the literature [35].

In recent years, machine learning (ML) approaches have been increasingly used to replace decisions that were previously made either stochastically or heuristically. For example, ML models can predict priorities between jobs instead of a priority rule [40], decide which variables are integrated in a constraint programming solver [41], or make decisions about neighborhood operators and acceptance decisions in a local search metaheuristic [42]. In addition, ML models are trained to be hyperheuristics that choose between different solution methods, with the goal of identifying the most suitable method for a particular given problem instance [43]. The common objective behind enhancing algorithms with ML-based decisions is that the benefit of an increased quality of the learned decisions outweighs the longer computation time required to query the models. The computation time required for training is often spent offline, so it has little impact on the final computation time required during inference. Regarding the implementation effort and the interpretability of the final algorithm, ML models introduce an additional layer of complexity that can negatively impact both.

In most cases, the ML models must make sequential decisions. This makes reinforcement learning (RL), which is designed for solving sequential decision-making problems, a popular choice. The background of RL, and of deep RL in particular, is given in the following subchapter.

2.2 Core Concepts of Deep Reinforcement Learning (DRL)

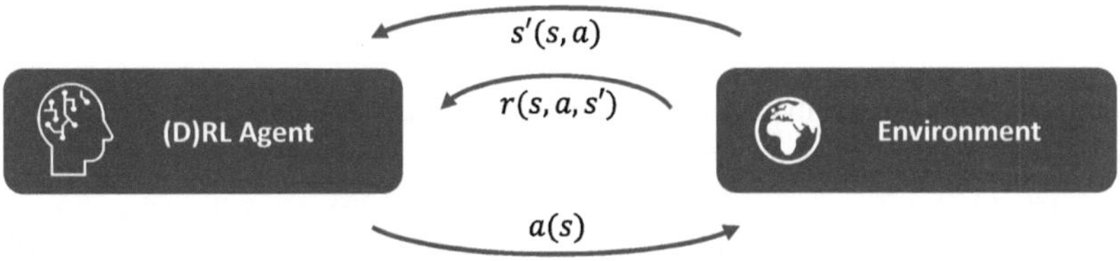

Fig. 2.5 Interaction pattern of a (D)RL agent and its environment

RL is a machine learning paradigm that aims to derive goal-directed solution strategies for sequential decision-making problems through a trial-and-error-like interaction of a so-called RL agent with an environment [44]. This environment comprises a certain problem and an evaluation of solutions to that problem [44]. The following descriptions follow those in Sutton and Barto's textbook on RL [44]. The sequential decision-making problem is described as a Markov decision process (MDP), typically formalized by the tuple (S, A, P, R, γ). In this tuple, S is the set of all possible states of the environment. A is the set of all possible actions an agent can take. P is the (probabilistic) transition function describing the probability of transitioning from state s to the subsequent state $s\prime$ following action a in state s. R is the reward function providing numerical feedback (rewards) r, for transitioning from state s to state $s\prime$. The discount factor γ defines how much future rewards are considered relative to more immediate rewards. The representation of the environment state provided to the agent is called observation. The interaction pattern between an agent and its environment is depicted in Fig. 2.5.

To ensure the theoretical convergence guarantees of RL, the decision-making problem must satisfy the Markov property, i.e., any current state is independent of all preceding states and actions that led to it. The goal of RL is to find a policy $\pi(a|s)$ that returns an action given a state that maximizes the expected discounted cumulative reward, called return G, of an episode. An episode is the trajectory from any initial state $s_{initial}$ to the final terminal state s_{final}:

$$G_s = \sum_{t=s_{initial}}^{s_{final}} \gamma^t r_{t+1} \tag{2.8}$$

Given a policy, one can attribute values to states, referred to as state-values V_s, , that are equal to the expected (E) return when being in said state and following the policy from there onwards. This can be written as follows:

$$V_\pi(s) = E_\pi(G|s) \tag{2.9}$$

Similarly, one can attribute values to actions in a certain state, that are referred to as state-action-values, or Q-values:

$$Q_\pi(s, a) = E_\pi(G|S, a) \tag{2.10}$$

If valid Q-values of all state-action-pairs are known, they can be used to derive an optimal policy, i.e., the one that maximizes the expected discounted cumulative reward, by evaluating all state-action-pairs in a current state and selecting the one with the highest Q-value. Thus, one way to perform effective RL is to learn valid Q-values from experience data in the form of state-action-reward tuples. This variant of RL is called Q-learning and belongs to the greater class of value-based RL methods. The recursive connection between states, rewards, and state values is formulated in the Bellman equation [45]. One of its manifestations is as follows: Given a state-value function, the value of a state at timestep t can be expressed as the reward for transitioning between state s_t and the next state s_{t+1}, plus the discounted value of the next state:

$$V_\pi(s_t) = E_\pi(r_t + \gamma V_\pi(s_{t+1})) \tag{2.11}$$

The same logic holds for Q-values:

$$Q_\pi(s_t, a_t) = E_\pi(r_t + \gamma Q_\pi(s_{t+1}, a_{t+1}))$$

This connection facilitates the iterative propagation of rewards in collected data through sequential states and updating the Q-values in a data-driven manner until convergence. A traditional way to store Q-values is in tabular form. However, tabular representations can only capture discrete state and action spaces, leading to the need for very large tables and leading to the limitation that if any state has not been seen in training, there are no Q-values on which to base the next action selection during inference. Due to these disadvantages of tabular Q-learning, value functions are often represented by approximation functions that can generalize to unknown inputs (states). The currently most popular choice for such functions are deep NNs [46], giving rise to the term *deep reinforcement learning* (DRL), to the extent that DRL and RL are often used interchangeably. The research field and the number of DRL algorithms are too large to be comprehensively covered in this thesis. Consequently, the following passages only introduce a taxonomy and briefly summarize the main intuitions of different algorithms that are relevant later in this thesis.

The company OpenAI has introduced a rough taxonomy of modern (D)RL algorithms which is shown in a slightly shortened form in Fig. 2.6 [47]. The broadest categorization, model-free versus model-based RL algorithms, are distinguished by whether the agent can make explicit use of a model of the environment (model-based) or not (model-free). In model-based RL, an agent is either given the model, or it learns the model, where the model consists of a transition function between states and a reward function for such transitions. A model can help the agent to plan ahead in a tree search-like fashion, which can be very helpful in highly uncertain, stochastic environments. The combination of Q-learning and MCTS facilitated by a model has produced the famous AlphaZero algorithm which achieved superhuman performance in Chess, Shogi, and Go in 2018 [1]. However, using a model for planning can also create computational overhead, which slows down computational speed of the learning and inference process. Since the JSSP addressed in this thesis is deterministic, model-free approaches are more relevant here and further explanations of model-based algorithms are omitted.

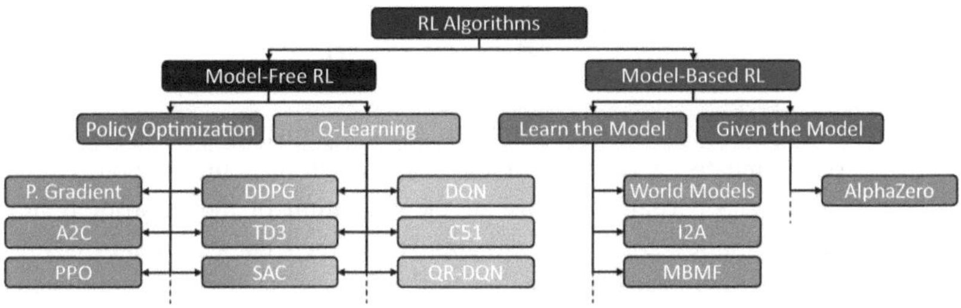

Fig. 2.6 Taxonomy of DRL algorithms (adapted from [48])

Model-free RL algorithms can be categorized into Q-learning algorithms (more generally referred to as value-based algorithms) and policy optimization algorithms (also referred to as policy-gradient-based algorithms). The goal of Q-learning algorithms is to approximate Q-values by a so-called Deep Q-Network (DQN) accurately enough to base successful decisions on them. A distinction can be made between algorithms that make use of data from complete episodes (Monte Carlo methods) and those that work with parts of episodes (Temporal Difference methods). Several key challenges arise during the Q-learning process. One challenge is that as the policy changes, previous Q-values may become invalid, potentially causing oscillations in the learning process. Another typical challenge for all DRL algorithms lies in the exploration-exploitation tradeoff. This tradeoff describes the critical balance between following a learned strategy when gathering new experience data to learn from, which is necessary for convergence but hinders the exploration of different strategies, and not following a learned strategy. Several tricks have been developed to mitigate such challenges, the form and combination of which make up the different algorithms in the Q-learning based algorithm family. For example, Double DQN mitigates a phenomenon in DQNs known as overestimation, where the Q-values are predicted to be too large, by introducing a second DQN and always using the smaller prediction of the two [48].

Policy optimization algorithms take a fundamentally different approach than value-based algorithms. They learn and update a policy network $\pi_\theta(a, s)$ parameterized by θ which, given a state s, directly returns a probability distribution over actions a that is interpreted as an evaluation of each action. According to one of the most popular Monte Carlo variants of policy gradient algorithms, REINFORCE, the parameters are updated by following rule [49]:

$$\theta \leftarrow \theta + \alpha \nabla_\theta \log \pi_\pi(s, a) G, \tag{2.12}$$

where intuitively, α is the learning rate to scale parameter updates, G is the return to scale updates according to the success of the current policy, and $\nabla_\theta \log\pi_\pi(s, a)$ is the partial derivative of the logarithm of the network policy outputs giving the policy-gradient-based approaches their name. A major advantage of policy-gradient-based approaches is the natural handling of continuous action spaces compared to value-based methods. A disadvantage is the instability of the REINFORCE algorithm resulting from the direct policy updates.

As with value-based methods, various techniques have been introduced over time to mitigate weaknesses of policy-gradient algorithms. Many of these techniques make use of value-function approximations to stabilize training, so that the distinction between policy-gradient-based, value-based algorithms, and the class of actor-critic methods that resides in between is highly debated and blurred. In this subchapter, the categorization by OpenAI serves as a point of reference [47], but in the remainder of this thesis a binary classification will be made for clarity, in which all algorithms deploying policy-gradient methods will be classified as policy-gradient-based and others as value-based.

Accordingly, notable examples of improved policy-gradient methods include Advantage Actor-Critic (A2C) and Asynchronous Advantage Actor-Critic (A3C) [50], Trust Region Policy Optimization (TRPO) [51], Proximal Policy Optimization (PPO) [52], and Option-Critic [53]. In A2C and A3C, the critic, i.e., the trained value function approximation network, is used to estimate the advantage of taking one action given a particular state over taking other actions, thereby providing indirect and smoother feedback that can be used for gradient updates. The terms "synchronous" and "asynchronous" refer to the data collection and update frequencies of different networks [50]. TRPO attempts to mitigate policy updates that are too large and lead to unstable behavior by minimizing the divergence between policies before and after updates [51]. PPO achieves a similar effect by introducing a clipped loss or a penalty term for large divergences between the policy distributions in the loss function, which is computationally more efficient [52]. The Option-Critic algorithm uses a hierarchical structure to choose between different options of sub-policies over temporally extended courses of actions [53].

A relevant representative of algorithms that are classified as actor-critic methods by OpenAI in [47] is the Deep Deterministic Policy Gradient (DDPG) algorithm [54]. DDPG addresses the limitation of value-based algorithms to work only with discrete and not continuous action spaces, but in contrast to previous algorithms employs a deterministic policy to generate the experience data.

There are two machine learning approaches that are related to DRL, because they also solve sequential decision-making problems, that are relevant in this thesis: Neural Evolution Strategies (neural ES) [55], and Self-Labelling (or Self-Improvement) [56, 57]. In neural ES, NNs are used to parameterize an action function. However, instead of training the networks with RL, the network parameters are stochastically perturbed to generate the training dataset before updating the parameters with stochastic gradient descent. This is done along gradients estimated numerically through the perturbed parameters instead of using analytical backpropagation [55]. In Self-Labelling (or Self-Improvement), multiple trajectories are sampled stochastically from an actor network. The actions taken in the best trajectory are then used as fixed labels to train the actor network in a supervised manner [56].

2.3 DRL for Job Shop Scheduling (JSS)

JSS can be formulated as an MDP and thus be solved with DRL. To illustrate how this could be done, an exemplary design of such an MDP is given in the following:

- State: The environment could hold the information of the complete JSSP instance to be solved, as well as information about the current (partial) solution to that instance, similar to the description of a partial Gantt chart.
- Observation: The environment could extract certain information from this state, such as the processing times of all next operations per job that have not yet been scheduled. Action: Based on this observation, the agent could prioritize between all jobs, so that the action vector contains the proposed job priorities.
- Transition: The environment would interpret this vector and integrate the corresponding operation as early in the current Gantt chart as possible under the given constraints.
- Next observation and reward: The environment could then return the next observation as well as the reward. The reward could consist of the difference of the intermediate makespans of the partial solutions before and after the latest action, forming a negative learning signal.
- Episode: This iterative process would continue until all operations of the problem instance have been integrated into the Gantt chart, forming a complete solution.
- Discounting: Since the environment is deterministic, later actions can be rewarded without a discounting factor ($\gamma = 1$).

This rough MDP design for the JSSP is just one of many possible ones. Designing the MDP, along with other algorithmic and logical design decisions, is central to the success of the training and inference protocols for DRL-based JSS solutions. In the following section, the necessary design decisions that constitute a method for building such solutions are outlined in a more generalizable way to build a framework for discussions of design choices made in the literature and in this thesis.

2.3.1 Necessary Design Decisions for DRL-based JSS

The necessary design decisions that define a DRL-based solution method for the JSSP are referenced in Fig. 2.7 and briefly explained below. Various choices for these elements that can be found in the literature are presented in Chap. 3. Note that some useful intuitions are derived from the literature in that later chapter but are already mentioned here.

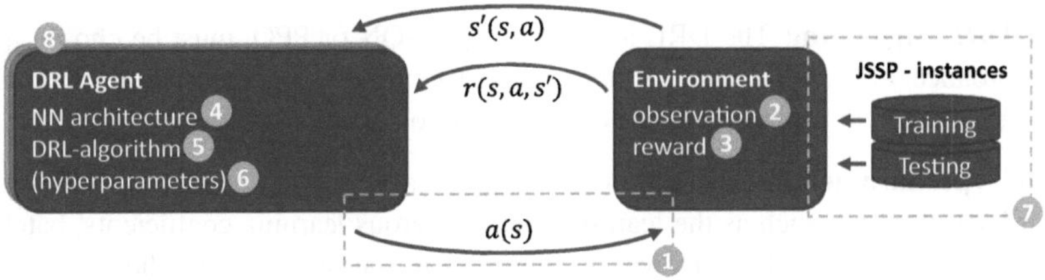

Fig. 2.7 Overview of the necessary design decisions of DRL solutions

(1) Action Space: The action space defines the semantic actions that the agent can take, and how the environment interprets them to construct or manipulate the current partial solution of the JSSP. Actions do not necessarily correspond to single scheduling decisions. One action could also consist of multiple sequential scheduling decisions, such as providing one prioritization that is used to schedule multiple jobs.

(2) Observation Space: The observation space defines the basis on which the DRL agent makes scheduling decisions. It should therefore contain information relevant to the strategy to be learned, such as information about the JSSP instance and the current state of the partial solution. The scope and expressiveness of features in the observation space determine how easily a problem can be learned. In general, it is easier to converge to a (simple) strategy for solving a task under restricted information, but the upper bound of the achievable solution strategy is more limited if essential information is missing [40].

(3) Reward: The reward is the learning signal for the agent and should relate to the makespan, yielding more positive cumulative rewards for achieving shorter makespans. Since the final makespan cannot be determined from individual actions until the schedule has been completed, engineered immediate rewards per action can introduce unwanted biases. However, sparse rewards that return only a final evaluation of the completed schedule can make the learning signal too weak to learn from [58, 59].

(4) NN Architecture: The NN architecture for policy and value networks can have a large impact on the resulting scheduling strategy. Some architectures may lend themselves better than other to represent the structure of the JSSP learning problem and to extract meaningful representations of the problem instance and intermediate solution [19]. However, if the architectures become too large, it can hinder effective parameter updates [60].

(5) DRL Algorithm: The DRL algorithm, e.g., DQN or PPO, must be chosen in accordance with the structure of the action space design, since not all algorithms are applicable to discrete and/or continuous action spaces [44].

(6) Hyperparameters: There are many hyperparameters that need to be adjusted for DRL training, such as the learning rate, numerous learning coefficients, batch sizes, or buffer sizes [61]. However, compared to the other design choices, such as the action logic and reward design, hyperparameters are less substantially different but only vary numerically, making them easier to optimize automatically.

(7) Training and Testing Instances: Depending on the desired outcome and the achievable level of generalization, one must define which, how many, and how often training instances should be used for learning. This can range from training on a single training instance to obtain a solution for it (no generalization) to training on many thousands of different training instances to learn a generalizing behavior that enables the agent to solve unseen instances without further training.

(8) Single- and Multi-Agent Approaches: DRL for JSS not only works with single agents interacting with the environment, but the underlying MDP can also be formulated as a multi-agent MDP. In such a scenario, individual agents may for example take the perspective of a machine, for example, and prioritize among the jobs waiting to be processed on it. Different agents may receive different local observations, individual rewards, and be otherwise constituted differently from others, which is why this design choice has an influence on all other design choices.

2.3.2 The Challenges and Vision of DRL-based JSS

The many design possibilities that exist within each of the design choices discussed above, and their mutual influence on each other, offer a large space for exploration and innovation in the research area surrounding DRL-based JSS. Although DRL is a good fit for solving the JSSP in theory, since the problem can be formulated as a classical MDP, no DRL-based solution strategy achieves consistently optimal results across all problems yet (cf. Chap. 3). The JSSP seems to pose a continuing challenge, and one can only make educated guesses based on the characteristics of the JSSP as to what makes the JSSP so challenging.

One reason could be the large but limited expressiveness of the NN architectures that are used, which have to capture and evaluate the entire, exponentially large solution space of a JSSP instance [2]. Furthermore, from this representation, an agent must learn how to reach the optimal solution through the interaction with the environment. This can be difficult, because a small difference in the order of operations on machines can have a large impact on the overall solution due to the combinatorial nature of the problem.

Another method-specific challenge is grounded in the difficulty of the agent to evaluate, if a solution strategy is optimal. This is the result of the inherent solution symmetry (cf. Sect. 2.1.2), which can cause many different action sequences to result in the same makespan, making the reward-based feedback signal noisy.

One challenge that could especially arise for agents that are supposed to learn strategies that generalize well to any given problem instance, is that there could be different strategies that work well in certain regions of the problem space. An agent may then easily converge towards some regions, but problems in other regions of the problem space may be optimally solved with a contraire strategy. Given such a phenomenon, it could be difficult to evaluate which strategy to use given a new instance.

Despite these difficulties the JSSP poses when addressed as DRL problem, significant progress has been made in recent years, as the review of the state of the art in Chap. 3 will demonstrate. With this progress on the JSSP in its standardized deterministic form, work on DRL-based JSS can be seen as contributing to a larger vision of end-to-end real-world DRL-based machine scheduling [62]. The path to this vision is schematically depicted in Fig. 2.8. As shown, the difficulty of the machine scheduling problems can be described by two broad dimensions: the complexity and accuracy of the representation of the factory that serves as the environment of a DRL-based scheduling solution, and the relevance of the evaluation criterion. In Fig. 2.8, these dimensions increase as one moves from the center to either side. The dimensions considered in the JSSP, i.e., machines, jobs consisting of ordered operations, and the makespan as evaluation metric, are close to the center.

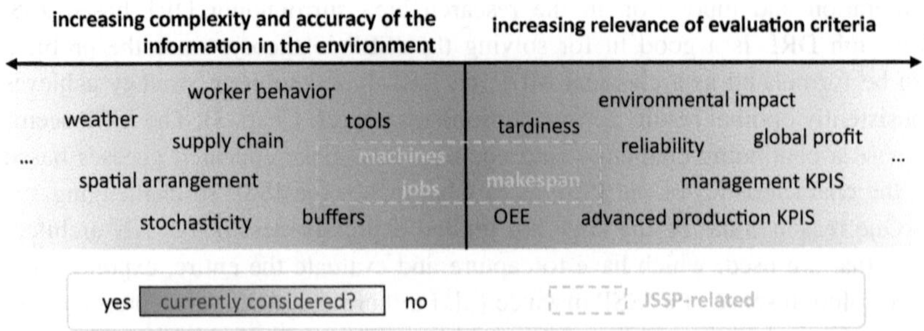

Fig. 2.8 Vision of end-to-end DRL based machine scheduling

In the real world, human workers often participate in manufacturing operations, meaning that individual capabilities and performance differences must be taken into account [63]. In addition, processing times can fluctuate stochastically, as can the error-free production rates of machines. Even factors that lie outside of the controllable environment, such as the weather, can have an impact on the productivity of workers or influence process parameters due to changes in humidity and temperature [64]. On the other hand, minimizing the makespan or maximizing the Overall Equipment Effectiveness (OEE) are arguably only local optimization criteria of a single shop floor within a company. They only indirectly serve the overarching goal of profit optimization. As explained in Sect. 2.1.1, machine scheduling is usually done in the context of other tasks on other strategic levels (Production Planning and PPC) with higher-level goals that the JSSP does not address. The vision is that DRL agents can automatically incorporate a larger amount of raw relevant information (one end) automatically and consider maximally useful evaluation criteria (the other end) to form an end-to-end, high-quality solution. While DRL-based methods are not yet very competitive with other existing methods on the strict JSSP, problems with increased complexity arising from the increasing accuracy of real-world problems and increasingly relevant evaluation criteria may be an area where the strengths of DRL-based problem solving make it a favorable solution method. In particular, potential strengths are the implicit consideration of stochasticity in environments and the reduced manual effort required to model relationships and constraints in the environment as logical simulations rather than mathematical constraints. However, solving JSS better is a necessary step for understanding and addressing the problem-specific challenges more effectively, before adding layers of difficulty to the learning problem.

3 Review of the State of the Art

This chapter provides an overview of the state of the art in DRL-based machine scheduling by means of a structured literature review. It serves as a connecting link between the preliminaries in Chap. 2 and the advancements on DRL-based JSS proposed in this thesis that are described in the subsequent Chaps. 4–7. As such, the overarching goal of the literature review is to provide the necessary context for understanding how the research questions and contributions discussed in later chapters relate to the current research on DRL-based machine scheduling. To achieve this, a representative literature corpus is generated that includes scientific publications on DRL-based JSS as well as relevant publications on DRL-based methods for machine scheduling more generally. The methodological process of the literature corpus generation is described in Subchapter 3.1. The resulting corpus is then analyzed in Subchapter 3.2, which is divided into three sections.

In the first section, Sect. 3.2.1, the entire literature corpus is analyzed with respect to two aspects: recent developments regarding which scheduling problems are addressed, and developments regarding the scientific contributions that are made in individual studies. As mentioned before, this thesis aims at advancing DRL-based JSS to contribute to the achievement of the larger vision: end-to-end machine scheduling solutions for real-world applications by means of DRL. Therefore, the analysis in Subchapter 3.2 distinguishes between two subgroups within the literature corpus: the first subgroup focuses specifically on JSPs ($(Jm| * |*)$ in three-field notation, meaning some additional constraints and objectives diverge from the standard JSSP). The second subgroup goes beyond JSPs and covers further related machine scheduling problems. The analysis of

C. Waubert de Puiseau, *Advancing Training and Inference Methods for Reinforcement Learning-Based Job Shop Scheduling*,
https://doi.org/10.1007/978-3-658-51305-4_3

this subgroup is valuable for understanding the broader scientific landscape in which research on DRL-based JSS is situated. The corresponding publications are listed in Table 3.1, in which the first row lists the publications of the first subgroup, and the remaining rows those of the second subgroup. Through this distinction, potential differences between the respective research communities working on the different problems can be identified. Depending on how similar or different the research communities are in terms of their background and contributions, there could be similar or different research gaps, trends, or preexisting domain knowledge. This can have consequences for the relevance of the results and contributions described in this thesis regarding the applicability to research questions addressed by each research community. Therefore, the results of the analysis in Sect. 3.2.1 will provide valuable information for the later discussion of the contributions regarding their applicability and transferability to other studies. This discussion is found in Chap. 8. Apart from the comparison between the two subgroups in the literature corpus and their respective research communities, the analysis in Sect. 3.2.1 sheds light on the following guiding questions: how rapidly is the research field expanding? Which research domains are involved in the research? And which research gaps exist? This information not only substantiates the general motivation for the research conducted in this thesis but also reveals potential blind spots for improvements that are currently overlooked and will be addressed in later chapters.

Subsequently, the focus of the analysis concentrates on those studies with a focus on JSPs, previously referred to as the first subgroup. Section 3.2.2, analyzes these studies and their corresponding suggested DRL designs with respect to the performance achieved by trained agents. This serves to gain a better understanding of what trained state-of-the-art agents can already achieve. Therefore, this section answers the guiding question of how the performance of an agent can be evaluated. A second guiding question is how well the performance of different agents can be compared. The motivation behind answering this question is the identification of top performing algorithms within the existing literature that can serve as an orientation point for successful design decisions.

Section 3.2.3 analyzes the statistics of design decisions made in the publications of the same first subgroup of literature on JSPs. The analyzed design decisions correspond to those described in the previously in Sect. 2.3.1. The goal of this part of the analysis is the identification of trends within certain design decisions made in the literature as well as the identification of common design choices made to design the top performing algorithms. Here, the underlying guiding question is which design decisions are already established, which are commonly experimented with, and what their influence on performance is.

Answers to these questions not only help to foster the intuition about design decisions that are promising candidates for future research and future applications but also help to situate the models used in this thesis within the space of possible design decisions made in the related literature. Finally, Sect. 3.3 provides a summary and reflection of the state of the art discussed in this chapter with respect to the guiding questions and the implications of the found answers for the contributions of this thesis.

As stated above, the goal of the literature review is to situate this thesis within the broader scientific landscape of DRL-based machine scheduling. As such, it covers a wide scope and intentionally leaves the level of detail necessary to form a comprehensive understanding for individual studies in this thesis to their dedicated literature sections. Therefore, additional complementary literature analyses are found in later chapters and subchapters, where recent related work is presented specifically for the corresponding research questions and studies. Related work on bottlenecks in JSSPs is presented in subchapter 4.1. Related work on curriculum learning in DRL-based JSS is analyzed in Subchapter 5. 1. Section 6.1.2 introduces the related work regarding action sampling in DRL-based JSS. Section 6.2.1 examines the related work on decision transformers for JSS. And finally, relevant literature regarding reliability in machine scheduling and its implications for DRL-based scheduling is analyzed in Subchapter 7.2.

The review in this chapter is based on the same literature corpus, poses similar research questions, and draws similar conclusions as a literature review that is currently in submission. However, compared to the content presented in this thesis, this separate article puts less emphasis on the JSSP but targets the broader machine scheduling domain instead.

3.1 Literature Corpus Generation

The literature corpus for the analysis in this chapter intentionally contains both subgroups mentioned above, of which the first focuses on JSPs and the second covers a wider range of machine scheduling problems. A multi-step search and screening process was used to build the entire literature corpus and is depicted in Fig. 3.1. The goal was to create a corpus size of between 100 to 200 publications, a size that is large enough to study statistics and trends over time but small enough to still be manageable for a manual full-text analysis. This was achieved through a rather restrictive keyword-based search, followed by a forward-search that aimed at identifying missing papers of high relevance, as indicated by citations from other sources.

In the first step, a keyword-based search was conducted on February 25th 2025 in three prominent academic literature databases: Scopus, IEEE Xplore, and ScienceDirect, and in Google Scholar. ArXiv preprints were explicitly excluded from the initial search, because they can lack a quality-assuring peer review process. However, they were included in the subsequent forward search, as their citation by fellow researchers is considered as approval by peers. In the following description of the search string, AND is defined as a logical "and", OR is defined as an inclusive logical "or", and the asterisk "*" represents a wildcard operator. The search included publications between the years 2017 and 2025, because no relevant literature using deep NNs was identified before 2017 and this constraint excluded unwanted publications using tabular RL. The initial search comprised publications with the terms "scheduling" AND "reinforcement learning" in the title, AND "job shop" OR "machine" OR "production" in their abstract.

Since this search alone returned more than 3000 publications, publications of related research fields that lie outside of the scope of the review, like network scheduling or computation core process scheduling, were identified and excluded by "AND NOT" operations through suitable keywords. Specifically, publications were automatically excluded, if their title, abstract, or keywords included "*grid", "cluster*", "cloud", "vehicle", "AGV", "maintenance planning", "maintenance scheduling", "bee", "genetic*", "high-performance computing", "photovoltaic", "solar", "PV", "suppl*", "network", "swarm*", "agricultur*", "FPGA", "VLSI", "evolution*", "robotic*", "compil*", "*core", "edge*", or "iot". Also, publications only available in Chinese language were excluded. The initial search yielded 381 results, from which 66 duplicates could be removed.

Next, titles and abstracts were systematically screened, and publications were removed if they met one of the following exclusion criteria:

- Being a review paper or concept/framework paper, because only papers with original content should be analyzed. None of the identified review papers covered the exact scope of interest regarding the literature corpus, and none reviewed the studies along all dimensions analyzed in this chapter;
- not addressing a production system;
- focusing only on location assignments;
- or primarily addressing robotic control.

Through this rough screening, 143 publications were removed, leaving 172 remaining publications. During the full-text analysis, further papers could be excluded and at the same time, relevant cited papers could be collected in a forward search. 37 publications were excluded, either:

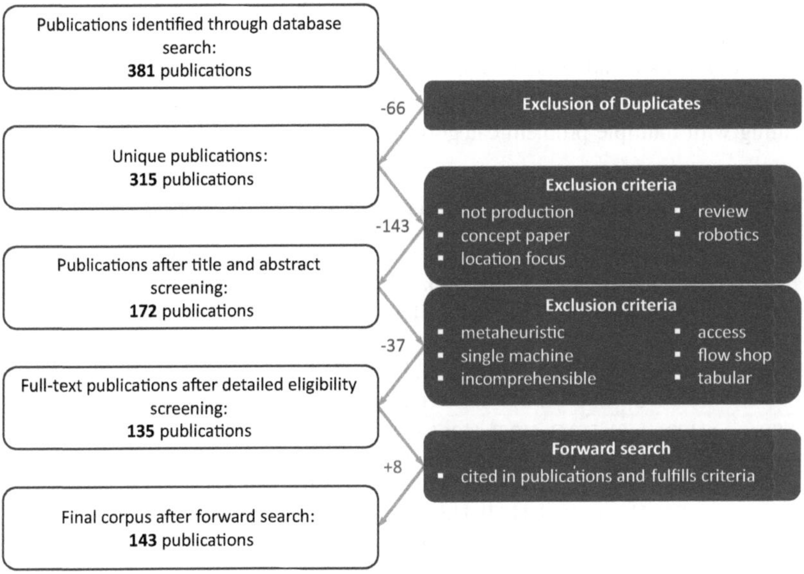

Fig. 3.1 Multi-step process for the literature corpus generation

- because the full texts were not accessible,
- or because tabular RL was used instead of DRL,
- or because DRL was only used to make high-level decisions in a metaheuristic,
- or because the quality of the paper was such that the content was incomprehensible, leaving out substantial parts of necessary information to understand the method.

In addition, single-machine scheduling problems and vanilla flow-shop scheduling problems were excluded, because they present a much easier problem and learning task, so that insights on these problems may be less transferable to others. However, problems dealing with considerable extensions of flow-shop scheduling problems through additional constraints were not excluded. Eight relevant papers were added in consequence of the forward search, resulting in a final number of 143 eligible publications. Although the resulting corpus cannot be claimed to be comprehensive, it falls within the previously defined range of 100 to 200 publications.

As a further consequence of the literature corpus size, the review in the following subchapter does not discuss every publication in detail. Hence, not every publication is cited individually. For reference, the resulting publications are listed in Table 3.1, sorted by problem type into which they were categorized. Papers dealing with multiple problem categories are sorted into all of these.

Table 3.1 Publications analyzed in the literature review sorted by problem category

JSP	[65], [66], [67], [41], [68], [69], [70], [71], [72], [73], [74], [75], [76], [77], [59], [78], [40], [79], [80], [81], [82], [83], [84], [85], [86], [87], [88], [89], [90], [91], [92], [93], [94], [95], [96], [58], [97], [98], [99], [100], [101], [102], [103]
DJSP	[104], [105], [106], [107], [108], [109], [110], [111], [112], [113], [114], [115], [116], [117], [118], [119], [96], [120], [121], [122]
FJSP	[123], [68], [124], [125], [126], [72], [16], [127], [128], [129], [130], [131], [132], [133], [134], [135], [136], [86], [137], [138], [139], [140], [141], [142], [143], [144], [145], [146], [147], [148], [149], [150], [151], [152], [153], [154], [155], [156], [157], [158], [159], [160]
DFJSP	[161], [162], [163], [164], [165], [166], [167], [168], [169], [170], [171], [172], [173], [174], [175], [176], [177], [178], [179], [180], [157], [181]
Parallel Machine	[182], [183], [184], [185], [186], [187], [188]
Flow Shop	[189], [190], [191], [192]
Other	[193], [7], [194], [195], [196], [197], [198], [199], [200]

3.2 Results of the Review

The results of the review presented in this subchapter are divided into an analysis of the broader metadata of the whole literature corpus in Sect. 2.3.2 (all publications in Table 3.1), and a subsequent more detailed analysis of the first subgroup of literature focusing on JSPs (first line of Table 3.1) in Sect. 2.3.1.

3.2.1 Coverage of Job Shops and Related Scheduling Problems in the DRL Literature

The field of DRL-based machine scheduling is rapidly expanding. The number of research articles in the literature corpus that were published in this field between 2018 and 2025 is shown at the bottom of Fig. 3.11. While only a few publications from the years 2018 and 2019 were found, the numbers increase steadily from twelve publications in 2020 to 37 publications in 2023. Although the data shows only 33 publications on the topic in 2024, which is a relative decrease compared to the previous year, this is likely just an artifact of conducting the literature search in February of 2025, when some recent publications from 2024 had not yet been indexed in the databases searched. One would expect the trend of the increasingly fast-growing body of literature to continue.

In Fig. 3.2, the absolute numbers per year are further subdivided and grouped into the number of publications per year and per problem category with which the publications primarily deal. Averaged over all years, studies on the static JSP and static FJSP are equally represented, accounting for 28.8% and 29.0% of all studies, respectively. Similarly, their dynamic counterparts, the DJSP and DFJSP, account for 13.7% and 15.1%, respectively.

However, a slight trend can be observed over time. While until 2021 the JSP and DJSP were the predominantly addressed problem settings, this trend changed in 2022 and 2023, and more recently, studies addressing the FJSP and DFJSP are found more often than those addressing the JSP or DJSP. At the same time, the ratio between the JSSP and DJSP has slightly but gradually shifted from a higher proportion of static to dynamic problems that are addressed in the literature. This trend towards dynamic problem settings in the JSSP is not clearly mirrored in the FJSP and DFJSP. Nevertheless, the general shifts from the JSP to the FJSP and from static to dynamic settings can both be interpreted as shifts towards problem categories that pose more difficult learning problems for DRL agents, since additional actions are required for flexible problems and uncertainties need to be learned in dynamic settings. This trend is likely attributable to the increasing quality of solutions that are achieved by DRL on the JSSP over time which will be shown later. Consequently, the community is gaining more confidence in applying DRL-based approaches to more difficult problems. This interpretation is promising, as it points in the direction of successfully pursuing the vision behind real-world DRL-based machine scheduling (cf. Sect. 2.3.2) over time.

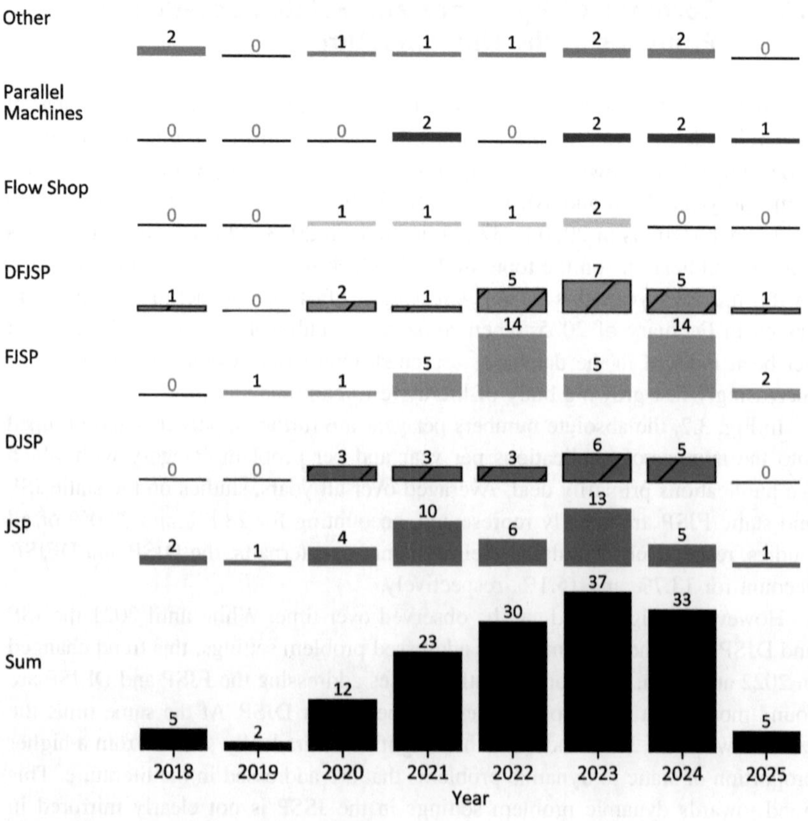

Fig. 3.2 Number of publications on DRL-based machine scheduling between 2018 and 2025

In the next step, the composition of the research community is analyzed to identify which perspectives on scheduling problems and solution methods are considered. This was done by looking at the most frequently used outlets for publications, i.e., journals and conference proceedings. For this purpose, the outlets are categorized into the following classes according to their dominant focus and audience:

- **Manufacturing**: This category features outlets with a focus on engineering tasks and corresponding technologies in the context of work in factories. Their titles often include terms like "manufacturing", "industrial", "engineering", "applied", "machines" or "production". Representative example outlets include the *Journal of Intelligent Manufacturing* or the *Conference on Production Systems and Logistics (CPSL).*

- **Operations Research and Computer Science**: This category features outlets with a focus on rather abstract planning and scheduling tasks and often more general solution methods for these tasks, or outlets with a focus on applied information technology. Representative example outlets include the *European Journal of Operational Research* or the *International Conference on Control, Decision and Information Technologies (CoDIT).*
- **Artificial Intelligence**: This category features outlets with a clear focus on advances in basic artificial intelligence technologies. Representative example outlets include the *IEEE Transactions on Neural Networks and Learning Systems* or the *International Conference on Learning Representations (ICLR).*

The categories are not always perfectly and objectively separable. As discriminating characteristics, the following ideas are considered in the context of the machine scheduling literature: The difference between "manufacturing" and "operations research and computer science" lies primarily in the motivation for studies. The former field focuses more on solving real-world scheduling problems, or abstracted versions as intermediate steps to solving real-world problems, and adapting existing solution methods to fit the problem. The latter focuses more on work that extends the toolbox of algorithms and technologies for planning and scheduling tasks, and is relatively more solution-method-centered than problem-solving-centered. The category "artificial intelligence" goes a step further. Here, the outlets focus on emerging AI technologies that can be applied to a variety of tasks. Machine scheduling is viewed rather as one of several examples to showcase the effectiveness of a new method than as the primary motivation of the studies. Two publications that resulted from the forward search were exclusively published as preprints on arXiv [104, 159]. Since the outlet arXiv covers preprints from various research fields, these two publications were both individually categorized based on their content, resulting in their placement into the "Operations Research and Computer Science" category.

All results of the categorization by outlet are presented in Fig. 3.3. The figure lists all journals and conferences in which at least two articles were published and shows the number of publications as a bar chart. All outlets in which only one article was published in literature corpus are summarized under "other". The colors of the bars indicate the introduced categorization of the outlets. One can see that outlets from all three categories are used by the community. In total, 81 different outlets were chosen for publishing 143 articles, indicating a wide variety of research fields that make scientific contributions to the interdisciplinary field of DRL-based scheduling. In general, articles published in conference proceedings account for 35.0% of all publications, and the majority, 65.0%, are published in journals. No journals stand out as having a larger number of relevant studies in this research field, indicating that it is still relatively young so that no specialized journal has yet emerged. Within the conferences, the Conference on Production Systems and Logistics (CPSL) stands out with four publications in the domain. This is a large number considering the small size of the conference (81 publications in 2024 [201]) compared to venues like the NeurIPS (4500 publications in 2024 [202]).

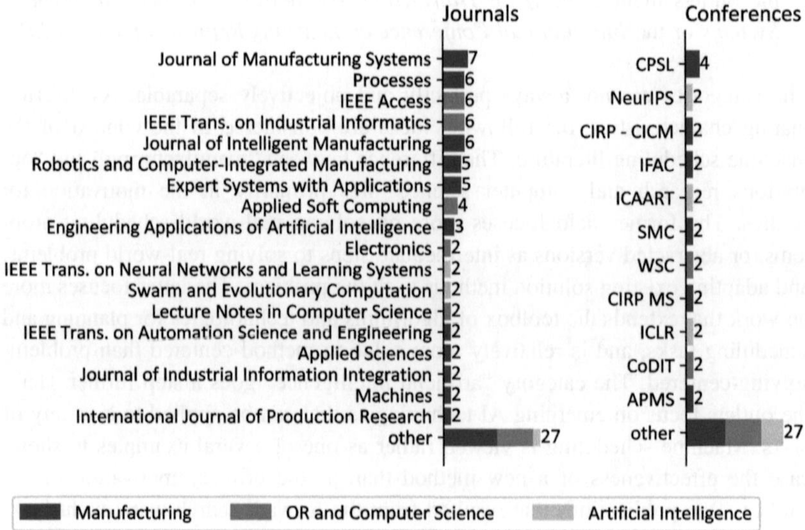

Fig. 3.3 Scientific outlets of all publications in the literature review

Beyond specific outlets, it is interesting to analyze which problem types are primarily addressed by which community, as it can hint towards potential strengths or blind spots that could emerge in the studies around specific problem types. Distinct communities can be represented by the respective outlet categories in which they choose to publish their results. Figure 3.4 shows which ratios of publications dealing with each scheduling problem category are published by which outlet category. The figure confirms the previous impression that most studies are published in outlets focused on manufacturing and shows that this is true across all problem categories. This visualization also reveals that while about 15–26% of the articles addressing the classical JSP and FJSP are published in AI-focused outlets, no articles dealing with their dynamic counterparts are.

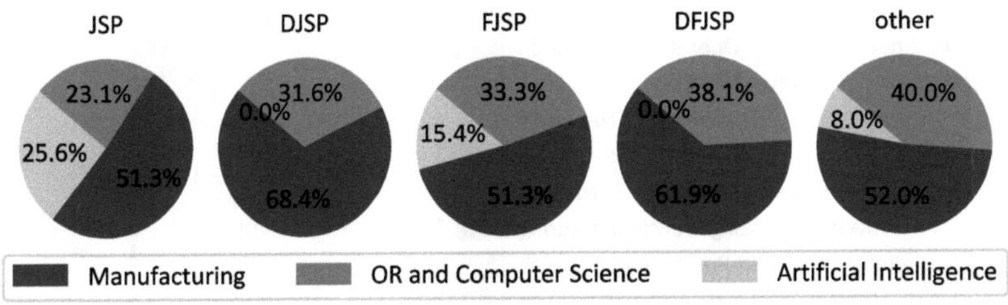

Fig. 3.4 Ratios of the numbers of publications per outlet focus sorted by scheduling problem class

There are two speculative explanations for this phenomenon. The first explanation results from the subjective observation that AI-focused outlets often require proof of superiority over state-of-the-art methods on public benchmarks. Therefore, it may be difficult to publish in such outlets, either because currently no universally recognized benchmark datasets and algorithms exist to easily compare against, or because DRL-based solutions are not yet as good as other methods in dynamic settings. A second explanation would be that the AI community is less interested in the problem settings to which the algorithms are applied but more interested in insights and methods that generalize to other domains, and therefore default to the most established problems in the literature. This explanation is also supported by the fact that the JSP, being the historically most studied problem, is the one that appears in AI-focused outlets most often. However, this could mean that there exists a research gap in applying the latest AI methods to more complex problems and, vice versa, a gap in applying domain knowledge to methods that deal with more abstracted problems.

The corresponding plot of the number of publications per outlet from the first subgroup of the literature, dealing with JSPs, is shown in Fig. 3.5. The figure confirms the previous impression that work in on DRL-based JSS is a multidisciplinary endeavor. However, this visualization of the JSP subgroup differs from the previous one that included both subgroups in several ways. Unlike before, in the JSP subgroup, the majority of articles (56.4%) is published via conferences. It is also evident that many articles are published at AI conferences with a top

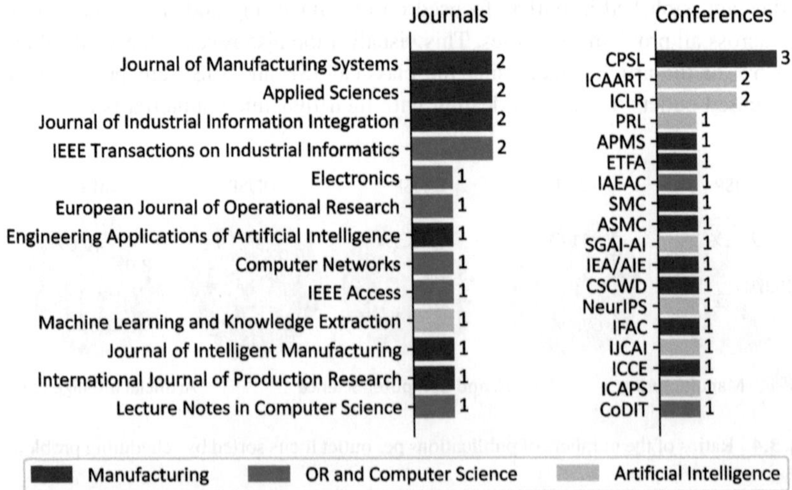

Fig. 3.5 Scientific outlets of publications on the JSSP in the literature review

core ranking [203], such as the A*-ranked ICLR, NeurIPS, IJCAI, and ICAPS, and the B-ranked ICAART. This indicates that the JSSP is a popular testbed for fundamentally new AI developments.

Upon closer examination of the JSP literature, a distinct emerging research field was identified, termed neural combinatorial optimization (NCO). Within the framework presented in this subchapter, the NCO community can be placed at the intersection between OR and AI. Distinguishing characteristics of NCO studies are that modern AI trends are quickly adapted and that the stated goal is the creation of general solvers for combinatorial problems. The JSSP is often only one of several combinatorial problems that are addressed in these studies by the same method.

As a final element of the meta-analysis, the contributions presented in the JSP-specific articles are analyzed to see where the current focus of development lies and where potential research gaps exist. To this end, the main contributions of each article are either collected from explicit contribution statements or inferred from the respective study. It should be noted that only because an article is not counted as providing a certain contribution, it could still do so. For example, almost all studies introduce a new variation of the observation space, but only few studies consider the observation space design as their main contribution or conduct ablation studies that reveal the importance of their particular design choice.

Therefore, the following results are only indicative of the areas where researchers subjectively saw the most room for improvement or what they considered to be the most important part of their work. Seven major directions of contributions were identified, most of which relate directly to the necessary design decisions presented in Sect. 2.3.1. These seven directions are: a novel NN architecture (NN Architecture), observation design (Observation), proof of concept that DRL can be applied to a certain problem (Proof of Concept, PoC), a novel problem extension to the JSSP (Problem) addressed with DRL, a novel reward function or study on the impact of reward designs (Reward), a novel environment design for the JSP or an MDP formulation for a JSP-related problem (Environment), or a new training algorithm or significant extension for DRL training on JSPs (DRL Alg. Training). A single article may make multiple main contributions. How often each main contribution was claimed is summarized in Fig. 3.6. According to these results, the most frequent major contribution is a new NN architecture, followed by novel environments, DRL training regimes and PoCs. The least frequent explicit novel contributions are made in the observation space. When the contributions are grouped into the three outlet categories defined above, certain differences within the groups become apparent. Whereas the distributions of the number of contributions in each direction are similar between *Operations Research and Computer Science* and *Manufacturing*, the *Artificial Intelligence* articles deviate from this pattern. Specifically, the latter domain makes fewer contributions through PoCs, new problems, new observations, and NN architectures. Instead, they seem to make more contributions regarding DRL training regimes and environments. It is intuitive that AI development would place less emphasis on new problems and PoCs, since it aims to make more generalizable contributions. Similarly, it is also intuitive that more contributions will be made on new training regimes and reward schemes.

However, three observations may at first seem counterintuitive: the strong focus on new NN architectures across all outlets; the small number of dedicated contributions regarding the observation space; the large number of contributions regarding the environment in AI outlets. These counterintuitive observations can be explained by a closer examination of the literature. While this examination will be done in more detail in Sect. 3.2.2, for the sake of completeness, certain insights regarding the counterintuitive observations are already given in the following analysis.

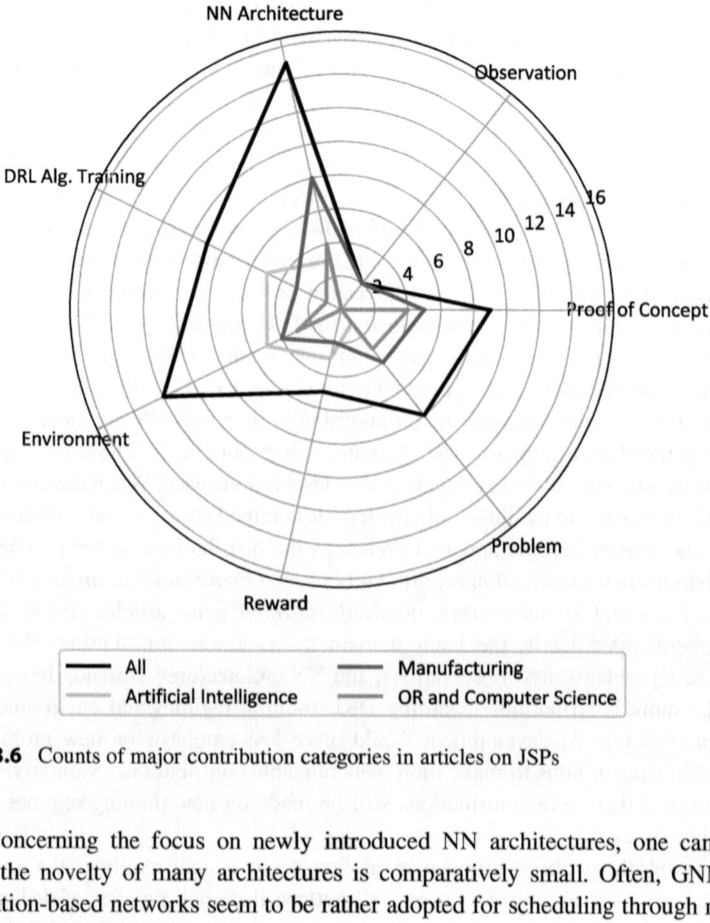

Fig. 3.6 Counts of major contribution categories in articles on JSPs

Concerning the focus on newly introduced NN architectures, one can find that the novelty of many architectures is comparatively small. Often, GNNs or attention-based networks seem to be rather adopted for scheduling through minor modifications than being newly developed. One can therefore argue that some contributions are not substantial in this respect. At the same time, in most cases, new observation spaces need to be developed for different architectures and applications. Therefore, the number of contributions of new observations should be larger. However, this effort is typically not reported as a major contribution. There are two possible explanations. Either the observation space design is intentionally not presented as a major contribution because it seems less novel and interesting than new NN architectures, or researchers tend to spend more time and effort on the development of NN architectures and therefore evaluate these efforts as a larger contribution. Finally, the emphasis on new environment designs in the AI literature can be explained by the different perspective that this community has on the JSSP. While environments are often designed to replicate online decision-making in a discrete event simulation fashion, the AI community is introducing other environments that construct schedules offline. Offline schemes offer more flexibility for integrating operations into partial schedules and therefore holds great potential.

In summary, the following conclusions can be drawn from the meta-analysis of the literature: DRL-based machine scheduling and JSS is a growing research field involving a diverse community of different disciplines. In recent years, there has been a gradual shift towards more complex problems that are addressed in the literature. Regarding studies that focus on JSPs, it seems that the observation space design remains underexplored, and that major contributions from the AI community target novel DRL training regimes and environments. These two findings support the motivation in this thesis to advance DRL-based JSS in terms of domain knowledge integration in the observation space and effective learning curricula, which are presented in Chap. 4 and Chap. 5, respectively. Furthermore, no study reported to focus on advancing algorithms in terms of the inference time, which is a fundamental performance metric as will be discussed in the next section, and which is addressed in Chap. 6 of this thesis.

3.2.2 Performance of State-of-the-Art Methods

Evaluating the performance, i.e., the achieved makespan in relation to the required computational time, of DRL-based JSS approaches introduced in the literature based only on reported results is not straightforward. However, being able to understand how well an algorithm performs in comparison to others is crucial for deriving the correct insights about design decisions from the comparatively large body of literature on DRL-based JSS. This section first elaborates on several dimensions that can have a large impact on the reader's ability to interpret the reported results. Given this context, the best performing algorithms in the state of the art are subsequently identified. This provides a foundation for the evaluation of prevalent design decisions that have proven successful, outlined in Sect. 3.2.3.

During the review of the literature, when attempting to compare the different studies with respect to their effectiveness in solving JSPs, six general dimensions emerged that, when they vary between studies, render the evaluation of reported results challenging or even impossible.

The first and easiest dimension to identify is the use of the same benchmark data and metrics for testing. The characteristics of public JSSP benchmark datasets were summarized in the previous Sect. 2.1.2. Figure 3.7 depicts the percentage of studies on DRL-based JSS report test results on one of the public JSSP benchmark datasets, and how often each dataset is used in the analyzed literature corpus. As can be seen, more than half, 22 out of 42 (56.4%), of all studies report results on public benchmarks. Within these benchmarks, the TA [22] and LA [25] datasets stand out as the most frequently used, followed by DMU [29], FT [24], and ORB [26]. Only one publication tested the model on the large-scale instances of the more recent DCT dataset [23].

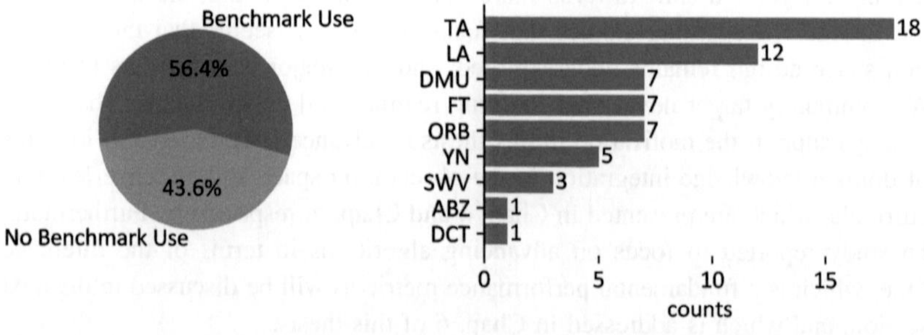

Fig. 3.7 Ratio of the usage of benchmark data in JSSP studies (left) and counts of usages of JSSP benchmark datasets (right)

Many studies report results on several public datasets. However, they often test either on those datasets containing smaller-scale instances, like LA [25] and FT [24], or on those containing small-scale to large scale instances, like TA [22] and DMU [29]. No study reports the results on all public benchmarks. Taking only the dimension of the used benchmark datasets into account, one might conclude that the results of 22 studies can be directly compared, because they all report their performance on the TA [22] dataset. Unfortunately, as already noted, there are more dimensions to consider.

A second important dimension is how much the designed algorithm must generalize when tested on the benchmarks. DRL agents are trained for different purposes, which is reflected in the degree to which they are required to generalize. In some studies, an agent may be trained on the same instance on which it is tested, leading to overfitting and the lack of generalization ability to unseen instances. Hence, the agent needs to be re-trained upon the introduction of a new scheduling instance. A more frequently observed goal level of generalization is achieved when an agent is trained on many problem instances of a particular size and then tested on different instances of the same size. A third level of generalization ability is expressed by training on instances of one size and testing on instances of a different size. None of the levels of generalization ability are superior to the others when they are fairly compared in terms of computational time required to generate the solution of a previously unseen instance. It can be posited that models that demonstrate superior generalization ability have been exposed to a more complex and challenging task. Such models may be perceived as more intriguing due to their ability to learn a heuristic that is widely applicable to the entire problem class encompassed by the training data, as opposed to a single instance. The main challenge for comparing performance, however, arises when studies do not make any train-test-split, i.e., they train on benchmark data instances and then test on the same instances during inference. Surprisingly, this was observed in 13/42 (31.7%) of the studies, impeding their fair comparison to other studies.

A third dimension that can hinder the comparability of algorithm designs is the manner in which results are reported. Unfortunately, there are many different ways to do so. Reporting the makespan on every instance in the dataset provides maximal information. However, for example, studies often report only the average results for all instances of the same problem size. Others report only the average makespan across all problem instances in the dataset or only the performance on a subset of instances per problem size. In addition, the results are not always presented as numerical values in tables, but sometimes only in figures, making it difficult to extract exact values.As a fourth dimension, some studies only report the result of generating one greedy solution, while other report the results of stochastically sampling multiple solutions from a trained policy (cf. Subchapter 6.1 for more details on sampling multiple solutions). This primarily hinders the assessment of the quality of the learned strategy. With respect to absolute performance, this means that the reported computational times for generating a solution must be considered for a fair comparison. This implication is related to two more dimensions that are crucial for the evaluation of reported results, because they all have a major impact on the computational time of each algorithm: The hardware specifications used in the experiments and the exact implementation. Often, information on these two details is not provided in the studies.

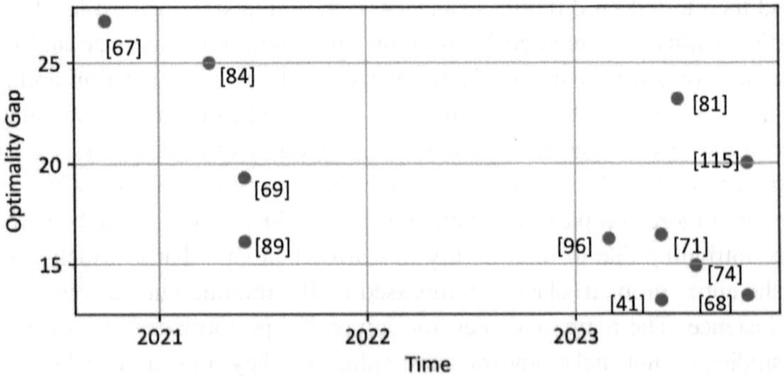

Fig. 3.8 Achieved average optimality gaps on the TA benchmark dataset over their publication date

Along these six dimensions, the 42 publications addressing JSPs can be filtered as follows: 22 out of 42 studies report results on the TA dataset [22]. Filtering out those studies that only report results on a subset of all TA instances [22] and those that strictly adhere to the one-factory setup, only eleven studies remain. These eleven studies are considered for direct comparison despite minor shortcomings. These minor shortcomings include:

- the best performing study which trained on all benchmark instances of the size 30 × 15 and reported generalized results on all problem sizes, including those on which it was trained, and additionally reported only sampled results but no greedy results [41];
- four studies in which the results could only be extracted graphically [69, 84, 96, 115].

The results were obtained in roughly comparable computational times on different, but always GPU-accelerated hardware. The results of the eleven studies [41, 67–69, 71, 74, 81, 84, 89, 96, 115] are plotted against the date of publication in Fig. 3.8. The bottom six publications are treated from here on as the most successful examples of DRL-based JSS algorithms, in order to support the assessment of useful design decisions in the following Subsection 3.2.3. It should, however, be noted that other publications not citing the TA benchmark [22] might still be informative, especially since the scheduling problems they address can diverge slightly from the vanilla JSSP formulation, e.g., by optimizing for machine utilization. It is therefore still worthwhile to consider them in statistics, even if their results cannot be compared exactly.

Compared to other constructive solution methods for the JSSP, such as those based on priority rules, DRL-based construction heuristics are competitive within a certain computational time window. However, NN inferences of the best models are much slower than priority rules. For example, Ho et al. [68] report an average of 82 seconds per instance, 3.5 seconds for the smallest and 444 seconds for the largest. Tassel et al. [41] report an average computation time of about 18 seconds per instance, which, assuming similar scaling effects, would equate to 0.7 seconds for the smallest instances and 100 seconds for the largest.

To put the current performance into perspective for the problem sizes in the TA benchmark dataset [22], depending on the environment implementation, priority rules take only milliseconds for small instances and a few seconds for the largest instances, and achieve optimality gaps around 19%. The latest version of OR-Tools [34], version 9.12.4544, achieves an optimality gap of 8.9% if granted one second of computation time per instance, and 4.9% if granted ten seconds. Thus, for the given problem sizes of most benchmark datasets, including TA [22], modern constraint programming solvers are far from being beaten by DRL-based solutions. However, as Tassel et al. [204] showed, DRL achieves better performance than other algorithms on very large problem instances, such as those in the DCT dataset [23], where solvers struggle to return any feasible solution. Hence, such problem sizes could be a realm in which DRL manifests itself as the best solution approach.

3.2.3 Common Designs and Design Trends

As discussed before, multiple choices have to be made when designing a DRL algorithm for JSS. This section analyzes the design decisions made for the study designs described in the literature corpus, with the goal of providing an overview of the many options that exist and identifying the most effective ones for optimal performance. Generally, there are three indications for effective design decisions: first, authors typically try out several designs during their research and design phase, and finally publish the overall algorithm they have converged on. Therefore, if a design decision is used more often in the literature, it is an indication that it works well. Second, as development continues, certain design decisions may become less common over time. Such trends can indicate that a newer design is preferable to the older one. Third, and most directly, the reported best performing algorithms strongly indicate that not only the individual design decisions but also the composition of several design decisions made in the respective study lead to good performance. For this purpose, the six best performing approaches defined in the previous section are used. To give the broadest overview of the state of the art for DRL-based JSS design decisions, all three indicative methods are applied to each design decision and the most informative conclusions are presented here. The design decisions analyzed in this way are, in that order: single- versus multi-agent systems, the action space design, the observation space design, the NN architecture design, the reward design, and the DRL algorithm used.

Single- versus Multi-Agent Systems

One of the fundamental decisions is whether to design a single-agent or a multi-agent system. In a single-agent scheduling system, one agent makes all scheduling decisions. Such designs are typically based on a global perspective facilitated by an observation space, containing information about all jobs and machines. In multi-agent systems for JSS, the perspectives of multiple agents representing only one particular machine or job are used, typically through a local observation. Fig. 3.9 depicts how often single- and multi-agent systems have been used in the literature over the years.

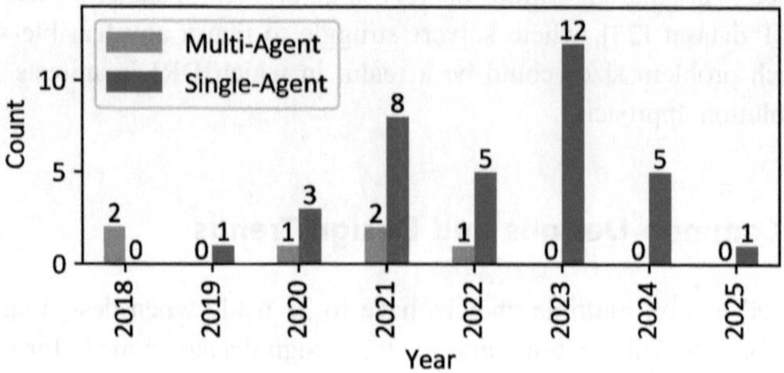

Fig. 3.9 Counts of single- and multi-agent systems introduced over the years

It is evident that the vast majority (85.4%) of designs choose a single-agent system and that no multi-agent systems have been introduced since 2023, showing a clear diminishing trend of multi-agent systems. In comparison, studies on problem classes other than JSPs show a much higher proportion of multi-agent systems (28.4%), as an analysis of the entire literature corpus revealed. This is because in these other problem settings, several different actions may be required, making the separation of different tasks for different agents more intuitive. Examples of such cases would be scenarios, where a choice must be made on which machine a job should be processed, or where different actions are required because multiple resource types exist, such as processing and transportation devices. For JSPs, on the other hand, the main argument for using multi-agent systems is that dividing the scheduling tasks among multiple agents makes the learning problem easier, especially for large problem sizes, because the observation space and the action space are smaller [65]. Examples for this include the work by Qu et al. [80], who provide local observations to agents that can only assign jobs to resources in their immediate vicinity, or the design by Waschneck et al. [7], which uses separate agents for separate work centers in a factory. However, with more expressive NN architectures and better ways to train them, this argument seems to have lost its significance over time. It should be noted that not every multi-agent setting necessarily trains multiple NNs. For example, Lohse et al. [65] use a hybrid approach by deploying a centralized critic network but multiple action networks. Park et al. [89], building one of the six best performing algorithms, take a multi-agent perspective with local observation and action spaces, but train a centralized policy with shared parameters. Notably, such an approach may, depending on the definition, still be classified as a single-agent approach. An interesting use case for a multi-agent approach is presented by Liao et al. [81], who deploy multiple agents in a hierarchical RL manner. Therein, one agent decides which of multiple specialized scheduling agents to deploy for the next time period.

Looking at the six best performing algorithms, almost all of them use single-agent scheduling schemes. The only exception mentioned above by Park et al. [89] can also be interpreted as a single-agent approach, since it trains a single NN. Overall, at least for the standard JSSP, it seems that approaches that deploy a single agent for making all scheduling decisions instead of separate agents per job or machine are advantageous. However, special cases like hierarchical RL, where multiple agents are used, but in which every agent takes global perspective, inherit all the advantageous properties of single-agent approaches and may be promising for future work.

Action Space Designs

The action space design, as defined in this thesis (cf. Sect. 2.3.1), has a great impact on the overall achievable performance of a DRL agent for scheduling tasks. As an integral part of the individual MDP formulation of JSPs, it is closely related to the environment design authors choose. In the following, the action space designs are characterized by four features: their action logic, how they handle action validity, single- vs. multi-decision designs, and direct vs. indirect designs.

The ***action logic*** defines the autonomy an agent has, to create a schedule or to manipulate a finished one. Six distinct action logics were identified in the JSP literature, which can be grouped into two categories: time-based and iteration-based scheduling logics. Figure 3.10 shows how often these action logics are applied in the literature. The action logics are explained below.

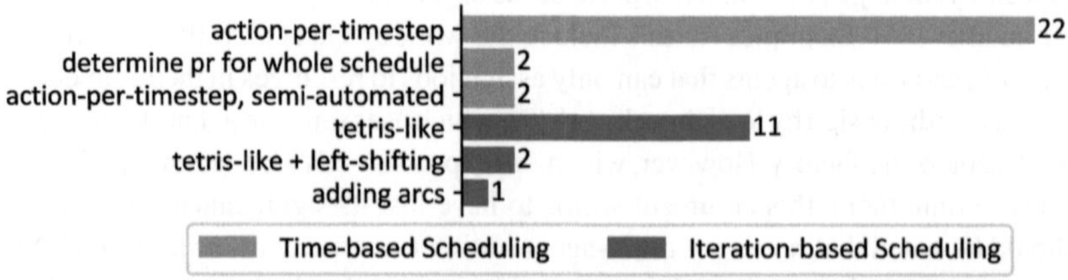

Fig. 3.10 Counts of action logics used in the JSSP literature

The difference between the two categories, time-based and iteration-based scheduling, is best described by their two most often used action logics: action-per-timestep and Tetris-like. Several logical iterations of both are depicted in Fig. 3.11. The logics are explained through their application to an exemplary small problem instance with three jobs, indicated by color, each with two operations. The required machine per operation is indicated by the index on the operation. In the action-per-timestep logic, a time-discretized process is adapted in which the agent is placed in a simulated situation that has a notion of the past, present, and future scheduling situation. In the situation depicted on the left, the simulation has stopped, because machine two has just finished. This marks an event, in which a decision (an action by the agent) might be required. In this situation, from the agent's point of view, machine one and machine three are running. Since no machine is currently idle and ready to process any remaining operation, no action can be taken and consequently, the simulation continues to the next time step without any action input. The next decision event occurs when machine one is finished (middle Gantt chart). Here, the next blue operation can be scheduled on machine one, which the agent initiates by taking the blue action. Then, the process continues.

The Tetris-like action logic functions differently. Here, the agent observes a partial schedule and chooses between all unfinished jobs of which the next operation should be integrated into the partial schedule. The chosen operation is then integrated as far left in the Gantt chart as possible without violating any constraints through the environment, similar to a Tetris block flying down as far as it can before hitting another block. Therefore, the agent has no notion of past, present and future, as it would be experienced while following a plan along the x-axis, i.e., the time axis. It has only a notion of past, present, and future iteration steps, not time steps, in the schedule construction process.

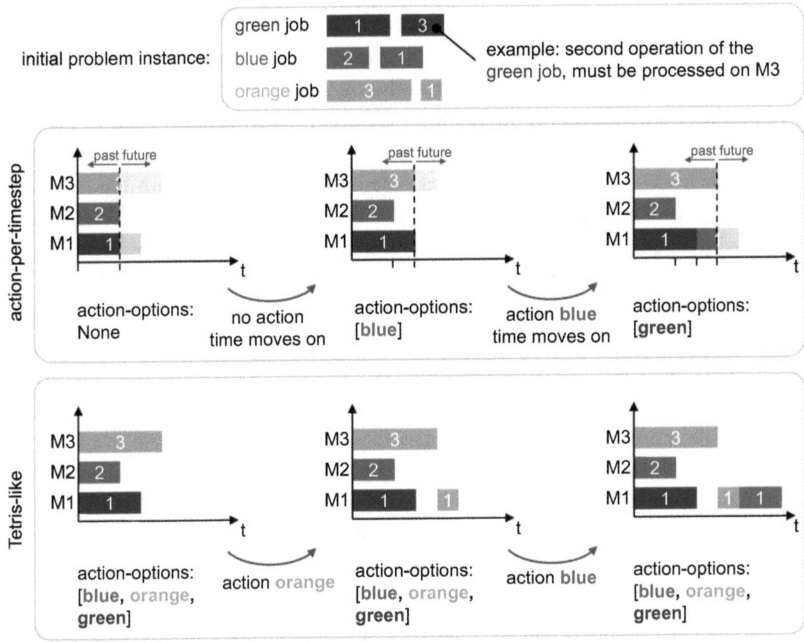

Fig. 3.11 Iteration steps of the action-per-timestep and Tetris-like action logics

The popularity of the action-per-timestep logic can be explained by four advantageous properties. First and foremost, moving along the timeline, and making decisions as they are required is intuitive to most practitioners who schedule reactively using prioritization rules. Second, this logic is applicable to discrete event simulations, which often form the basis for industrial use cases. These simulations are convenient to use as training environments because they implement all constraints, provide a visual interface, and often already have interfaces to resource management systems. Third, the logic is directly applicable to online decision-making settings for problems beyond the static JSSP, i.e., for problems with dynamic changes over time, such as machine breakdowns or new job arrivals. Fourth, since actions are taken whenever a machine becomes idle, large gaps in the schedule are often avoided. Consequently, even randomly taken actions result in schedules with acceptable makespans. Figure 3.11 illustrates this property, showing that the action-per-timestep logic results in a shorter schedule. However, the fourth advantage can also be a disadvantage. Taking actions whenever they become possible introduces a strong bias for the agent to schedule operations early. Therefore, the schedule seen in the third depicted iteration step of the Tetris-like action logic would be very unlikely or even impossible to produce with the action-per-timestep logic. Thus, the latter logic often cannot cover the entire solution space.

Iteration-based scheduling logics have different benefits. They tend to be less biased towards taking early actions, and have more actions available to them per step, which leads to greater exploration potential. Also, they do not have to deal with partially finished operations or currently running machines in their observation spaces, which generally keeps the observation spaces smaller and more efficient.

The other two time-based scheduling logics share many properties of the action-per-timestep logic. The "action-per-timestep, semiautomated" [70, 76, 96] logic follows the same process, but does not query the agent for a decision if only one option exists, as is the case in the example in Fig. 3.11. The "determine the priority rule (pr) for the whole schedule" [103] logic chooses one priority rule that is applied to the entire schedule but still has the character of a discrete event simulation. However, the agent does not learn to make every scheduling decision as a construction heuristic but learns to determine one good rule as global construction heuristic depending on the problem instance it observes. Instead of iterating over many individual scheduling decisions, it then iterates over many training instances by taking only one action per instance.

The other two iteration-based scheduling logics also share most characteristics with the Tetris-like action logic. The "Tetris-like + left-shifting" logic [67] differs in the way in which selected operations are integrated into the schedule. If switching the order of the preceding operation on the required machine with the chosen operation in this iteration step shortens the immediately achieved makespan, this switching procedure is applied. This procedure shows very good performance even when used in conjunction with priority rules as decision-making module, but it takes away some of the agent's autonomy and has the effect of decreasing the solution space [67]. The "adding-arcs action logic" behaves much like the "Tetris-like action logic" but is unique to the graph representation of the JSSP [92]. All initially undirected arcs are iteratively redefined as directed arcs by pointing at the direction they should attain. This is noticeably different from a normal construction heuristic, since, for example, it allows the order of operations on a single machine to be already defined, before the order of operations on other machines is. The environment is responsible for finding the shortest path on the resulting graph with (partially) directed edges in order to form intermediate schedules or the final schedule.

As Fig. 3.10 shows, the action-per-timestep and Tetris-like action logics are by far the most popular ones for the reasons already mentioned. Looking at the six best papers by benchmark results, all but one of them apply iteration-based action logics. The exception is one algorithm that determines one priority rule for the whole schedule [71]. However, this one performs worst among the six best algorithms. The results are therefore a strong indication that iteration-based logics are MDP formulations for the JSPs that are easier to learn and lead to better overall results.

The next discriminative criterion deals with the question of how to handle ***action validity***. As described above, invalid actions can occur in many action logics, when resources are not available or jobs are already finished. Two general approaches can be identified in the literature: learned and enforced action validity. As the name suggests, learned action validity refers to strategies in which taking invalid actions is possible but gets punished through negative rewards. Conversely, enforced action validity strategies refer to those in which taking invalid actions is made impossible. There are three general options to achieve this and effectively enforce valid actions. The most common one is action masking [205], where the value or action prediction vector of the DRL networks is set to $-\infty$ for the neurons representing invalid actions. The second option is to design a value network with only one output neuron, which is queried separately for the evaluation of each action at a time step. If there are invalid actions, the network is not queried for the evaluation and thus they are excluded from the set of possible choices. A third option is indirect actions, such as choosing a priority rule. In this case the environment handles the logic of eliminating invalid actions from the set of possible choices. This works by rule-based filtering for valid actions and prioritizing only between the resulting actions. Indirect actions are discussed in more detail later in this section.

Figure 3.12 shows both the overall share and the counts of the use of learned and enforced action validity in the literature over time. Studies using enforced action validity account for the majority (82.9%) and there is a relative decrease over time in the number of studies opting for learned validity designs. This is because teaching the validity of actions necessitates the introduction of an additional reward term, which not only complicates the engineering effort, but also the learning task, and does not guarantee that all actions are valid during the usage of the model. It is likely that enforced validity is the more popular option due to these disadvantages of learned validity. This is also reflected in the six best performing algorithms on the standard JSSP, all of which employ action masking. However, enforcing action validity may be impossible in problem settings where a sequence of actions leads to

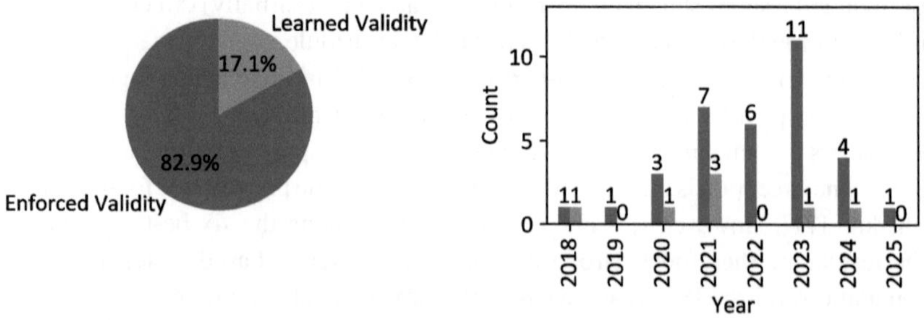

Fig. 3.12 Relative share (left) and counts over time (right) of learned and enforced action validity designs

a constraint violation, such as violating a production deadline. In such cases, reward engineering can still be necessary.

Another distinguishing feature between action space designs is using ***single- or multi-decision designs***. These designs refer to the number of decisions an agent must make to perform a scheduling action. Multi-decision designs are more common in flexible scheduling settings, where a job and a machine must be appointed to start the next operation. For JSPs, only three (3/42, 7.1%) multi-decision designs can be found. In one of these, Bonetta et al. [66] deploy a pointer network that defines an entire schedule in one action. Another example is introduced in the study by Kuhnle et al. [59], where a decision is represented by a start and an end resource. The agent thereby points at a queue of finished jobs on a source machine and simultaneously decides which job should be taken by choosing an appropriate next machine. Lastly, Liao et al. [92] use a hierarchical DRL system in which two distinct agents each take a single action, attributing to a multi-decision design in which one agent chooses between scheduling agents and the other, chosen scheduling agent makes a scheduling decision. However, single-decision designs largely dominate the literature on JSPs. And all six top performing designs rely on a single-action design, because outputting a single action generally constitutes an easier learning task than outputting multiple coordinated actions in parallel or sequentially.

Finally, a distinction can be made between ***direct and indirect action*** designs. Direct actions refer to designs where the agent directly references jobs or machines to be scheduled next. This grants the agent maximal control over the scheduling process. Conversely, in indirect actions, the agent chooses between prioritization rules or outputs a regression value as indirect prioritization, as done in [58]. The environment then interprets the action and performs the prioritization and scheduling. For example, a processing time may be predicted, and the environment chooses the operation whose processing time is closest to the output value to form a prioritization. Indirect actions facilitate the integration of action validity filters handled by the environment. They can also restrict possible actions to those that help to guarantee a minimal decision quality, for example, if only strong prioritization rules can be chosen from. On the other hand, indirect actions limit the explicit control of the agent over the decision which operations or resources are considered next. For JSPs, only 21.1% of all studies use indirect actions, while for the DJSP, arguably the more difficult learning task, 55.0% use indirect actions. This suggests a trade-off between facilitating good performance on difficult learning tasks and granting enough control for very good performance on easier learning tasks. Therefore, most of the currently best performing algorithms for the JSSP make use of direct action designs. One exception is the work by Ho et al. [71], where an agent chooses one priority rule for the whole schedule. However, their algorithm performs worst among the six best performing algorithms.

Observation Space Designs

Observation spaces form the basis on which agents make their scheduling decisions. They must represent the scheduling problem and the state of the current partial solution in a sufficiently detailed but efficient manner. Despite the simplicity of the raw problem description through three features per operation (the processing time, the job and the order within the job), the number of different features for observation spaces that can be found in the literature is large. They do include the already mentioned raw descriptions of operations as numerical features, but also simple descriptive features about the current state, such as booleans that indicate whether a machine is busy, an operation is currently running, or an operation is already finished. In addition, it is common to compute the number of remaining operations or the remaining processing times until the instance will be solved, for both jobs and machines.

All these features are consistent with the information that is used in common priority rules to estimate the immediate prioritization of jobs or machines, such as SPT or MTR (cf. Section 2.1.3). In fact, it has been shown that the scheduling strategy of the agent assimilates these rules given the corresponding information as observations [40]. However, there are also features whose correlation to a successful strategy is not intuitive. One example is the utilization rate of a machine in the solution up to the current state. Furthermore, some features involve more than the sum or average of values but rather the computation of the product or ratio of certain features. Finally, certain more elaborate problem descriptions involving job deadlines or weighting factors, machine breakdowns, or dynamic machine operation costs, will typically result in the use of features that describe these elements numerically. Typically, features are either min-max scaled to the interval [0, 1] or standardized within their feature type. Given the large number of different features, a statistical analysis of individual features is not useful. Instead, the higher-level distinctions between engineered and raw features as well as myopic and non-myopic observations are analyzed in the following.

A rough distinction between observation spaces can be made depending on whether they consist of ***engineered or raw features***. Raw features refer to those that describe the problem instance in a complete but minimalistic way. Engineered features are those that involve any calculations, i.e., taking the sum, average or similar operations, beyond scaling and standardization. Since almost all observations contain some raw features, all observation spaces containing at least one engineered feature are categorized as using engineered features, while the raw features category is reserved for those containing raw features exclusively. The exclusive use of raw features is representative of the vision of end-to-end DRL for scheduling problems, where minimal bias is introduced through human feature engineering. The overall share and counts of both categories over time in the literature are depicted in Fig. 3.13. Roughly one third of all observation designs rely on raw input features only. There is no clear trend towards more or less reliance on raw features over time and in the years with most included publications, 2021–2023, exactly one third of all observation designs were based on raw features. This suggests that the community is not actively moving towards the vision of end-to-end DRL in this respect.

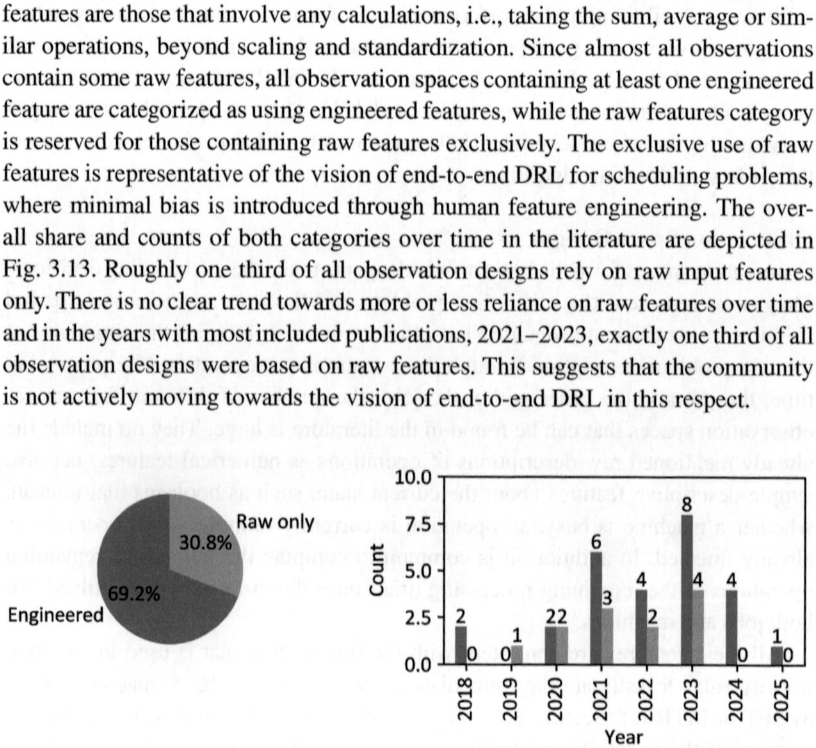

Fig. 3.13 Relative share (left) and counts over time (right) of observation spaces consisting of engineered or raw features

The second distinction between observation spaces is made between ***myopic and non-myopic*** observation spaces. Myopic observations only provide information on the next operations or machines, rather than those that will need to be considered in the more distant future. An example of a myopic perspective is the SPT priority rule, which only considers the processing time of the next operation per job and ignores information about machines and subsequent operations. Statistically, only 12.5% of all designs are myopic. A possible explanation is that, in principle, a myopic observation presents an easier learning task, but even the best strategy that is based on limited information will be far from optimal for many scheduling instances.

Conversely, providing full information is a more difficult learning problem, but the agent may struggle to effectively derive a strategy at all. Thus, there is a balance to be struck between providing enough information to enable the creation of a good scheduling strategy and providing less information to facilitate the convergence to a good strategy. Accordingly, most observation designs fall somewhere between myopic and full (non-myopic) information. Another option is the inclusion of full information and theoretically redundant more myopic features to facilitate training. The use of only myopic features was found predominantly in studies prior to 2021 [7, 77, 80, 85] with one exception in 2023 [98]. This trend can be attributed to the development of more stable overall algorithm designs featuring, for example, PPO, dense rewards, and stable hyperparameters. Such developments facilitate successful agent training with more complex observations.

Looking at the six best performing algorithms, five out of six utilize engineered, non-myopic features. These provide full information, e.g. via GNNs, but also provide theoretically redundant information via engineered features. One exception is the design by Tassel et al. [41], who provide raw features, not for all operations, but only for the next few operations per job. They showed that providing full information did not increase the final performance, but slowed down network inference, confirming the hypothesis that a balance between limited and full information may sometimes be required.

Neural Network Architectures

The choice of NN architecture plays a critical role in the performance of the final agent, because it has a major impact not only on the expressiveness of the policy, but also on the inference time. Furthermore, some architectures inherently express the structure of the scheduling problem and, once trained, can be applied to different problem sizes. Therefore, the choice of NN architectures can often be the determining factor for other design choices. The literature has been analyzed to identify dominant architectural elements used in overall NN designs. Most architectures contain at least one fully connected layer (FCN). To avoid redundancy, only designs that exclusively use FCN layers were classified as "FCN". For studies that do not provide details on their NN implementation, the use of FCN was assumed and labeled as "FCN (?)". Studies were also classified according to whether they include elements of convolutional neural networks (CNN), self-attention elements

(ATTN), recurrent NN structures (RNN) and graph NN properties (Graph-NN). Since one architecture can feature multiple of those elements at once, multi-class classification is possible. However, NNs that are primarily Graph-NN, but leverage convolutional or attention-based elements in their message-passing structure, are classified as Graph-NNs only. Figure 3.14 shows the number of occurrences of the NN elements in the JSSP literature.

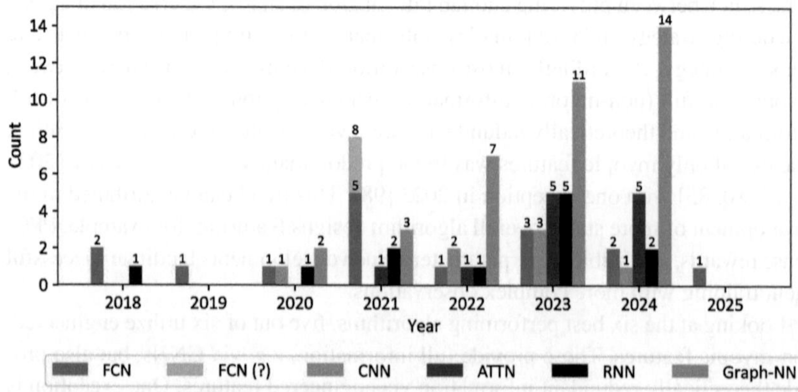

Fig. 3.14 Counts of artificial neural network elements featured in studies over the years

Early studies up to 2021 largely relied on FCN architectures. Although their relative share has been decreasing since then, they are still used in recent years. This circumstance is likely due to the fact that FCNs are often pre-implemented and default architectures in popular software frameworks such as StableBaselines3 [61] and are therefore readily available. While CNN, RNN and ATTN elements are becoming more popular, with ATTN elements showing the clearest trend, the usage of Graph-NNs has grown most steadily and significantly, being used in more than half of all NN architectures in 2024. GNNs were first applied in the scheduling domain by Zhang et al. [67] in 2020. Their popularity can be explained by the intuitive way in which Graph-NNs model scheduling problems and their constraints through structure. It is therefore not surprising that five of the six best performing architectures deploy Graph-NNs. The one exception is the NN designed by Tassel et al. [41], who use the ATTN mechanism. It is noteworthy that all these studies design size-invariant architectures and show good generalization ability from training on small problem instances and testing on larger ones.

Reward Designs

Reward designs determine how easily successful scheduling strategies can be attributed to individual actions. Thus, they have a significant impact on the speed of convergence and the learnability of scheduling problems. For statistical evaluation, one can broadly classify dense and sparse reward designs. Dense reward designs refer to those, in which a meaningful reward is given after every step and it would be conceivable that all these rewards would be non-zero. Sparse rewards are those, in which constant rewards, most often zeros, are given. Here, a true evaluation is only done sporadically or even only once per episode. How often both alternatives were used over time, as well as the number of cases in which both alternatives were featured, are shown in Figure 3.15.

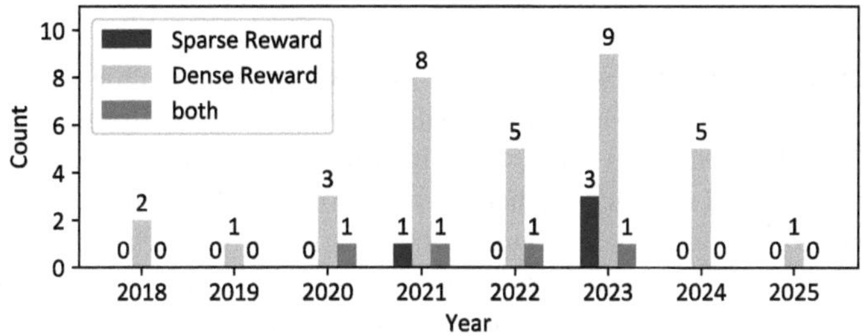

Fig. 3.15 Counts of dense and sparse reward designs over the years

Apparently, most studies use dense reward designs, with three exceptions using sparse rewards [41, 78, 96] and four exceptions doing both [59, 124, 161, 184]. Dense rewards are given such that either, the cumulative reward is inversely proportional to the makespan or to the cumulative machine utilization. This is done through a negative reward as large as increase in makespan that an action causes, or through a reward proportional to the machine utilization per timestep.

The three exceptional papers that use sparse reward design differ from the majority of the overall algorithm design, which motivates their choice. Tassel et al. [41] aim to prove that an end-to-end algorithm that can handle sparse rewards and to thereby pave the way for more complex performance metrics than the makespan. Similarly, Su et al. [96] lay emphasis on the advantages of evolution strategies over more classical DRL strategies and use sparse rewards according to this strategy. Pirnay et al. [78] leverage a self-competition design, where the superior agent is rewarded with a +1 for winning, relying on the evaluation of completed episodes through sparse rewards. Two papers combine a dense and a sparse reward design.

Liu et al. [85] designed a dense reward signal proportional to the processing time of the job selected in the last action and combined it with a final sparse reward term. The latter is proportional to the difference between the final makespan and a lower-bound makespan. Vivekanandan et al. [82] use a dense reward to train action validity. This dense reward is augmented with a sparse final reward proportional to the makespan, only if the agent reaches a final solution within a fixed number of steps.

Interestingly, two studies also directly compare sparse and dense reward designs in their study. The results in both studies indicate that dense reward signals lead to faster learning and convergence to comparable or better final performance [58, 59]. It is therefore reasonable to assume that in most JSP settings, dense reward designs are a good starting point when designing a new algorithm.

DRL Algorithm Choice

The last analyzed element is the choice of the DRL algorithm to train the NNs to form policy functions. A broad binary classification between value-based and policy-gradient-based algorithms is made, where hybrid algorithms such as actor-critic or DDPG are classified as policy-gradient-based. Evolution algorithms (cf. Sect. 2.2) are counted as a separate class. Figure 3.16 shows how often each DRL algorithm was used. The algorithms are colored by their binary class. "Other" summarizes algorithms that were used only once or were not described well enough to be identified as a specific algorithm beyond their binary class. For policy-gradient-based algorithms, single occurrences include TRPO, DDPG, A2C, A3C and Option-Critic. Within value-based algorithms, the one exception is an on-policy TD algorithm. Overall, one can see that policy-gradient-based algorithms are the predominantly used class of DRL algorithms. The literature also shows that the number of value-based methods is declining over time.

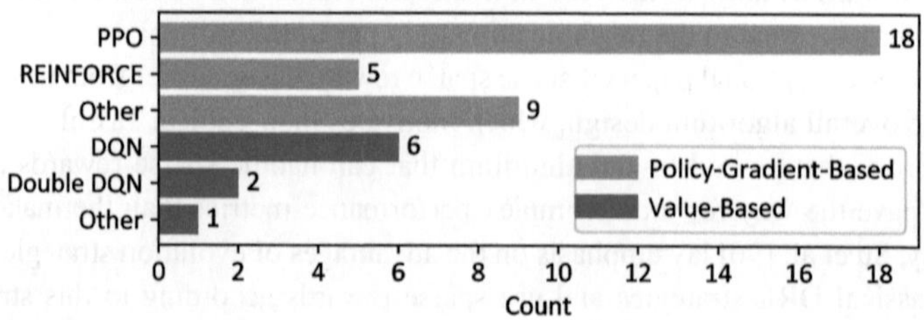

Fig. 3.16 Counts of used DRL algorithms sorted by policy-gradient-based and value-based algorithms

Among policy-gradient-based algorithms, PPO is by far the most often used. The stated advantages of PPO for JSPs include:

- stable learning behavior,
- a comparatively small sensitivity to hyperparameters,
- reported superior performance, as shown in direct comparison with SAC by Samsonov et al. [97], and to the vanilla policy gradient algorithm, A2C and A3C by Maharjan et al. [101].

PPO's applicability to both discrete and continuous action spaces and well-documented online implementations [61] may also contribute to its popularity. Evolution strategies have been applied only once.

Interestingly, a closer look at the six top-performing studies and the DRL algorithm they applied paints a slightly different picture. Ho et al. [71] successfully leveraged the value-based Double DQN algorithm in 2023, giving no explanation for their choice although more recent algorithms would also have been available. The design Iklassov et al. [74] is the only one within the top six that applies a classical PPO implementation, whereas Park et al. [89] use a related REINFORCE implementation with clipped loss, Tassel et al. [41] also use a related REINFORCE implementation, and Ho et al. [68] use classical REINFORCE in their "Residual Scheduling" design. Su et al. [96], as stated before, use evolution strategies. Thus, although the community seems to be converging towards using PPO, and it has shown superiority in direct comparison in certain cases, other algorithms may still be worth comparing. This is especially true for policy-gradient-based ones that integrate stabilizing features such as a clipped loss or penalty terms.

3.3 Reflection on the State of the Art

This chapter described a literature review on DRL-based JSS featuring a large literature corpus comprising 143 publications with a broader scope of DRL-based machine scheduling problems, not only JSPs. The analysis of this corpus revealed information about the constitution of the research community that is involved in this topic. A closer look at the subgroup of literature dealing particularly with JSPs allowed the evaluation of the performance of current methods, the identification of major contributions that are made in this domain, as well as trends regarding the decisions made for necessary design choices.

It was found that DRL-based JSS is a growing and highly interdisciplinary field, in which many different areas of expertise are combined into working algorithms. Three such areas of expertise were identified from the outlets through which the publications in the literature corpus were published: Manufacturing, Operations Research and Computer Science, and Artificial Intelligence. From this thesis's perspective, which aims to advance DRL-based JSS, inspiration can be drawn from different research communities. In fact, Chap. 4 aims to integrate domain knowledge from the Manufacturing and the Operations Research and Computer Science fields to produce answers for RQ1, how domain knowledge can be advantageously integrated into the observation space. Chapter 5 takes inspiration for new methods from the AI community to provide answers for RQ2, how training strategies should be designed for efficient and effective DRL agent training.

Regarding the analysis of the contributions that are made in the publications, it was found that a lot of focus is put on modifying and improving necessary design elements for the training process of DRL agents. As one would expect, the reported solutions are generally becoming more powerful over time, and consequently increasing effort is put into training models to solve increasingly difficult scheduling problems. While much progress has been made in training methods along the necessary design elements, little progress has been made to improve inference methods after training is completed. This thesis addresses this research gap and explores the option for advancing the field in this direction in Chap. 6 to fill the research gap. The chapter thus answers RQ3, how trained agents should be used during inference for optimal performance.

The literature review further provides an overview of the common design choices that are made as well as trends in these choices. Certain identified trends, such as the use of PPO or size independent NN architectures, stand out for their clarity and may therefore treated as recommendations for adapting these choices. At least, they can function as starting points when designing new solutions or as benchmark systems against which one may compare. For other design decisions, even in cases where the vast majority of studies make the same decision, for example opting for dense rewards, no clear recommendations can be made, because they might not be applicable to the rest of the design. Despite attempts to identify more general trends that lead to the success of the currently best performing algorithms, it is therefore difficult to map individual design decisions to their contribution to the performance improvements without ablation studies. In this thesis, this insight is taken into account when introducing novel methodological elements, by not applying them to entirely new algorithm designs, but by integrating them to existing well-performing algorithms instead. That way, any changes in the training behavior and final performance can be attributed to the novel element.

Finally, it should be stated that the literature review conducted in this chapter is not without limitations. A major limitation is the way in which the literature corpus was generated. The keyword-based search was rather restrictive, which may have left relevant publications unnoticed. However, in order to manage the large body of literature, a restriction had to be made. It was tried to mitigate this shortcoming through a forward search, which is useful for finding such publications that were deemed relevant by fellow researchers in the field. The initial search also explicitly excluded any studies on computation core process scheduling or network scheduling problems, two important adjacent research fields. These fields likely pursue slightly different scheduling problems, emphasize short computation times more strongly, and are therefore not directly comparable to machine scheduling solutions. However, in a future effort, it would be worthwhile exploring the advances made in these somewhat separate research areas, as the corresponding communities could provide yet alternative perspectives and novel ideas that are overlooked in the machine scheduling community.

Another limitation is that only those publications dealing with problems that are very closely related to the standard JSSP were analyzed in more detail. Hence, insights were primarily generated for DRL-based scheduling for this problem class and may not all be transferable to other problem classes. While the chosen scope targets the core of this thesis, it is not well aligned with the greater vision behind DRL for JSS in general (cf. Sect. 2.3.2). Therefore, in Chap. 7, a portion of the literature corpus is revisited to discuss trends with respect to real-world scheduling problems, which relates to RQ4.

The study shows that different studies report results differently, on different benchmark instances, and with different expected generalization levels needed of trained agents. Hence, even within the 42 publications that focus on JSPs, it is difficult to evaluate and compare the performance of the presented DRL-based scheduling solutions objectively. A level of difficulty is added when trying to attribute the performance of a trained agent to a single design choice, since many design choices work in unison to form a comprehensive solution. Therefore, the attempts to identify design choices that are favorable over others in the previous chapter are based on limited evidence for the effectiveness of individual design choices rather than hard facts. However, the identification of better or worse design choices is not the main goal of this chapter. Rather, it is now easier to conceptualize whether the contributions in this thesis can be seen as natural extensions adding to previous work, such as introducing a new observation design, or if they explore distinct new paths for improvements that are not the focus of previous work. Additionally, the insights that studies are difficult to compare and that improvements are difficult to attribute to certain design choices are a major motivation for the study designs in this thesis. To make the effect of algorithmic advancements more easily comparable to other studies, novel elements will mostly be added to preexisting third-party solutions.

Domain Knowledge Integration in the Observation Space

4

As discussed in previous chapters, the JSSP has been the subject of extensive study for decades, particularly within the OR community. As a result, the problem has been analyzed in depth, leading to specific characteristics and structures inherent to scheduling problems. Such characteristics have been put to practice in the design of tailored heuristic decisions in solution algorithms. In other words, much *domain knowledge* about the JSSP has been acquired in the OR community over the years and forms the basis of many tailored solutions. This makes them particularly effective for this problem class. The aim of this chapter is to explore how such domain knowledge can be systematically integrated into DRL-based JSS solutions to make them more effective, both in terms of convergence speed and final solution quality. As will be shown later, one natural way for integrating domain knowledge is through the observation space. Thereby, this chapter contributes towards answering RQ1, *how domain knowledge can be advantageously integrated into the observation space.*

The term *domain knowledge* can be defined as follows: "The realm of knowledge that individuals have about a particular field of study. As such, domain knowledge encompasses declarative (knowing that), procedural (knowing how), and conditional (knowing when and where) knowledge" [206]. In this thesis, it describes not only the knowledge of individuals but a whole research community. While screening the literature on classical solution methods for the JSSP, two reoccurring characteristics of the JSSP were frequently mentioned: the problem is highly solution symmetrical [21] and individual instances contain certain resource bottlenecks [207]. These two traits appear to be declarative knowledge on the JSSP. Moreover, the research community has adopted these traits and turned them into procedural knowledge, leveraging symmetries and bottlenecks algorithmically to enhance existing or design novel solution methods [9, 21].

C. Waubert de Puiseau, *Advancing Training and Inference Methods for Reinforcement Learning-Based Job Shop Scheduling*,
https://doi.org/10.1007/978-3-658-51305-4_4

Here, solution symmetry refers to the characteristic that numerous distinct solutions result in the same makespan. While solution symmetry will be discussed later, in Sect. 6.1, this chapter is built around the notion of bottlenecks in the JSSP.

The term *bottleneck* is colloquially used to describe the critical component within a system that constrains the rate at which a desired outcome can be achieved. This metaphor is derived from the narrowest part of a physical bottle, where the cross-section has the smallest diameter, is referred to as bottleneck and has a significant impact on the flow rate of a liquid from it. A query of abstracts containing the terms "scheduling" and "bottleneck" (connected with or without a logical "AND", published after 2010) in the scientific literature database Scopus in December 2024 suggests that 2.179 out of 171.385 (1.3 %) documents concerned with any scheduling problem are also concerned with bottlenecks. These numbers showcase the large scientific interest around the notion of bottlenecks in scheduling problems. In the corresponding literature, bottleneck *machines* seem to be the most frequently addressed limiting resource. The notion of bottleneck machines and other resources in process optimization has even found its way into popular literature, such as the bestseller novel "The Goal: A Process of Ongoing Improvement" by Eliyahu M. Goldratt from 1984 [208].

The extensive attention devoted to bottlenecks underscores their significance to the problem class and its solution, and therefore clearly qualifies as domain knowledge. The central premise underlying scheduling strategies that incorporate the notion of bottlenecks is that, by definition, bottleneck machines constrain the throughput or productivity of the whole system. Hence, they should be given special attention, ensuring their operation at maximum capacity to allow the system to operate efficiently. A common pattern emerges across scheduling strategies involving bottleneck machines, comprising a sequence of iterative steps. A first step (1), in which the bottleneck is identified from the current system state of jobs and machines, a strategy development step (2) that dictates how the bottleneck resource can be prioritized, and a strategy application step (3) that generates or alters the schedule accordingly. One iteration of these three steps is schematically depicted in the top section of Fig. 4.1. The bottom of the figure depicts major functions of a DRL iteration, namely (1) the observation generation, (2) NN inference and (3) action. Thereby, Fig. 4.1 illustrates one intuitional way in which parallels between traditional bottleneck-based scheduling algorithms and DRL algorithms can be drawn. Noteworthily, the strategy development step, in DRL, is a black-box algorithm learned by a NN. While alternative parallels could be drawn and are discussed in subchapter 4.4., the central idea in this chapter is motivated by the illustrated parallel and adheres to the following train of thought:

If a scheduling problem contains a bottleneck resource and an effective strategy exists to leverage this information, a DRL agent could be trained to learn such a strategy through interaction data. To facilitate the recognition of a bottleneck resource, one can provide explicit information about existing resources, regarding the degree to which they present bottlenecks, through features in the observation space. For convenience, any such features are referred to as *bottleneck information* from this point on. Borrowing from successful bottleneck identification schemes in the OR literature for the augmentation of the observation space is an intuitive and direct way to integrate domain knowledge into DRL-based solution strategies.

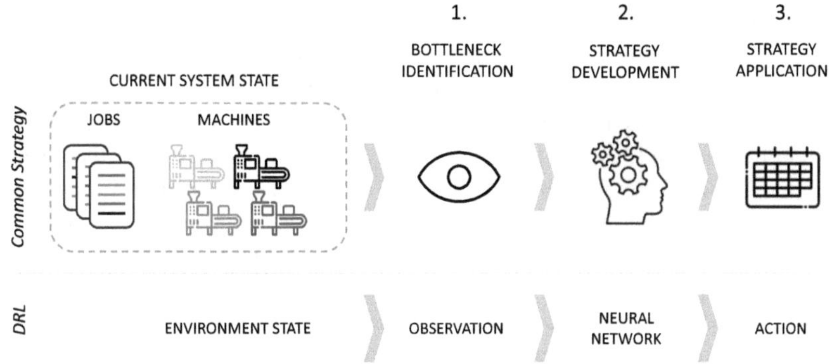

Fig. 4.1 Common steps of scheduling strategies under bottleneck considerations

Building upon this idea, this chapter explores the possibility of increasing the convergence speed and final performance through the addition of bottleneck information into the observation space as a candidate for effective domain knowledge integration into DRL-based solution methods. To this end, the chapter is structured as follows: first further related work is reviewed in subchapter 4.1, comprising bottlenecks metrics in the OR domain in 4.1.1 and notions of bottlenecks in the DRL-based JSS literature in 4.1.2. The reviews are aimed at examining the extent to which notions of bottlenecks have been integrated into DRL-based solution approaches and identifying promising candidate methods for quantifying bottleneck criteria for such purposes. In 4.1.2, the experiment design is described before the results are presented in 4.3, examining the influence of the inclusion of identified bottleneck information. The hypothesis of the study is that the inclusion of bottleneck information will increase the convergence speed and final performance of the DRL algorithm. The chapter is concluded with a summary and discussion of the contributions of this chapter with respect to RQ1 in 4.4. The content of this chapter expands the information published in [209], mainly through updated reference literature and a more in depth discussion of the results and the contributions in the broader context of this thesis.

4.1 Related Work: Bottlenecks in JSSPs

4.1.1 Bottleneck Metrics and the Shifting Bottleneck Heuristic

Many advanced heuristic techniques deploy bottleneck metrics for the identification and quantification of bottlenecks in production systems. For example, bottleneck machines are commonly identified by a large average utilization, maximum or average queue length in front of them, or the average waiting time of jobs in their queue [210, 211]. These metrics capture characteristics of the schedule that closely resemble the definition of bottlenecks as resources that limit the throughput of the system. The advantage of such bottleneck information is that they are cheap to compute and applicable in online scheduling scenarios, since only information on past events or, depending on the action logic, the already scheduled operations is needed. This is why such features have already been included in many observation space designs of the publications reviewed in Chap. 3. However, a moving average across past decisions and the resulting schedule are not necessarily indicative of the future. Especially in the vanilla JSSP, a machine which presented a bottleneck in the beginning of the problem, for example measured by a high utilization, may in fact not be a bottleneck for the remaining problem, since many of the parts it has to process are already completed by then.

The disadvantages of the above metrics are mitigated in simulation-based bottleneck information generation schemes. For example, Zhang and Wu identify limiting bottleneck machines and critical jobs through the statistical analysis of multiple forward simulations. They then use this information to steer different genetic algorithms [212–214]. The authors analyze similar information as described before but they are calculated for theoretical futures. This approach is similar to model-based DRL algorithms that do Monte-Carlo simulations to plan into the future, with the critical difference that bottlenecks are identified in rollouts instead of objective values. Although empirical evidence was presented showing that such bottleneck information is valuable, the identified bottlenecks depend on the scheduling strategy deployed in the forward simulation. If no strategy is used, i.e., schedules are stochastically generated, the combinatorial solution space is questionably too large to obtain useful bottleneck information for DRL agents in feasible time.

A widely known and highly cited heuristic specifically for the JSSP is the Shifting Bottleneck Heuristic (SBH) [9]. As the name suggests, the notion of bottlenecks is a central component of the heuristic. For a long time it has been the gold standard for the JSSP, and consequently, variations of it have been devised for related problems [215]. This makes the SBH an interesting candidate for the purpose of this study. To provide an intuition for the bottleneck identification and the strategy used to leverage this knowledge in the SBH, the algorithm is briefly summarized in the following paragraph according to the wording used in Pinedo [11]. For more details on the SBH, the interested reader is referred to the original publication [9].

The SBH repeatedly identifies the current bottleneck machine and determines a new order of all operations on that machine until the order of all operations is determined and the solution has converged to a final result. The procedure in each iteration is applicable to an unsolved JSSP, a partial schedule and a full schedule alike. This is achieved through the graph representation of the JSSP as introduced in 2.1.1. Recall that each node represents an operation in the JSSP instance with two additional dummy nodes for the start and end node. These nodes are connected by two types of directed edges which represent precedence constraints within jobs, and processing-time weighted precedencies on each machine, respectively. To determine the current bottleneck machine $m_{bottleneck}$ from the set of all machines M, , first, the longest path in the graph of the (partial) solution is calculated, which serves as a lower bound for the achievable makespan. The lower bound solution allows the determination of pseudo due dates for all operations that would have to be met in order not to increase the lower-bound makespan. The algorithm then iterates over all machines m_i to find pseudo release dates for all operations that need to be processed on m_i. The pseudo release dates result from minimizing the maximum tardiness objective in the surrogate graph representation, in which the edges of all machines but those of m_i are fixed. In other words, all operations on machines $\forall m(m \in M, m \neq m_i)$ are fixed and only the order of all operations on m_i is altered. Critically, this one-machine tardiness optimization problem is still strongly NP-hard. However, it can still be solved much more efficiently than the entire JSSP at once. After iterating over all machines, the machine with the largest tardiness t_i is considered the current bottleneck machine. The order of operations on $m_{bottleneck}$ is then set to the one found by minimizing the maximum tardiness in the bottleneck determination iterations. This concludes one iteration of the heuristic.

4.1.2 Bottlenecks in DRL-based JSS

Bottlenecks are not unknown in the DRL-based JSS literature either. In fact, there are 24 publications in the literature corpus that is analyzed in Chap. 3 that mention the term bottleneck. Seven of those explicitly refer to the SBH as a competitive solution alternative for the JSSP. However, they do not claim to draw any inspiration from it in their algorithm design [68, 69, 75, 82, 199, 216, 217]. Interestingly, the authors in thirteen publications recognize that their addressed production system could have bottleneck resources or refer to bottlenecks that are identified in the schedules resulting from their proposed method. Yet, this information is never used proactively to find better algorithms or solutions [6, 7, 58, 59, 107, 138, 192, 198, 218–222].

However, there are also four publications in which the authors make proactive use of the notion of bottlenecks in their algorithm design. Zhou et al. [106] recognize that an even distribution of workloads mitigates the formation of bottlenecks. Following this reasoning, the authors include the standard deviation of utilization rates in a partial schedule as a feature in the observation space. Following a similar reasoning, van Ekeris et al. [40] introduce an operation-wise so-called "bottleneck feature" by multiplying the processing time of each unscheduled operation with the sum of all remaining processing times of operations that require the same machine. Both methods incorporate the domain knowledge of the possibility for the existence of bottlenecks in their observation space design but deploy rather straightforward bottleneck metrics. Moreover, the impact of those specific features on the convergence rate and final solution quality is not analyzed.

Popper et al. [125] devise a different strategy based on a bottleneck-oriented reward design that biases the agent towards optimizing the tardiness of jobs on a pre-determined bottleneck machine. This is achieved by rewarding only the deviation of a target tardiness and the achieved tardiness on that resource. However, the publication lacks details on the procedure for bottleneck identification. Yet another distinct approach, called MINERVA, is introduced by Thomas et al. [161]. MINERVA features two separate NNs, one for the explicit identification of bottleneck resources in the system and one for Q-learning. The bottleneck identification NN is trained separately on experience data collected from simulations, using the current Q-learning DRL agent in different environments, in which the capacity properties of machines are varied. If the throughput increases after the capacity of a machine has been increased in simulation, it is labeled as the bottleneck resource of the system. The bottleneck identification NN is trained to predict these labels based on the features "average weighting time", "average queue length", and "average utilization" of each resource over a fixed time interval. Importantly, the Q-learning DRL agent decides on the jobs to schedule next and on the resource with which the job should be processed. However, the agent has no explicit information on the current bottleneck, hindering dynamic strategy adjustments to account for dynamic bottleneck formations. Rather, the bottleneck identification is used to increase the capacity of a resource in the system. In contrast, in the study presented in this chapter, the agent is given explicit information about the current bottleneck via features in the observation space.

4.2 Experiment Design

The experiment is designed to analyze the influence of the integration of bottleneck information, specifically a metric adapted from the SBH, on the convergence speed and final performance of DRL designs for solving the JSSP. To this end, this subchapter describes the calculation of an SBH-inspired bottleneck information feature, two different DRL base models and how they are augmented with the bottleneck information, and the training procedures.

4.2.1 Bottleneck Feature

The bottleneck information calculation procedure is closely inspired by the bottleneck identification step of the original SBH [9]. However, it contains minor modifications to be applicable to intermediate solutions of construction heuristics. The pseudo-code is presented in Algorithm 1. *Algorithm 1: Bottleneck Information Calculation Pseudo-Code*

Algorithm 1: Bottleneck Information Calculation

```
input: current_schedule;
bottleneck_info_list ← [ ]
makespan_LB ← calculate lower-bound makespan
due_dates ← calculate pseudo due dates
for every machine do
    tardiness_MIN ← minimize tardiness of one-machine-problem(current_schedule, machine,
                    due_dates)
    bottleneck_info ← makespan_LB + tardiness_MIN
    bottleneck_info_list.append(bottleneck_info)
end
min-max-scale(bottleneck_info_list)
output: bottleneck_info_list
```

The input to the function is the current schedule of a construction heuristic, e.g., any partial schedule including a completely empty schedule at step zero, represented as graph of the entire problem instance with partially directed edges (cf. Sect. 2.1.2). It should be noted that this is different from the SBH which only deals with partial solutions in which all or none of the directed arcs of operations on the same machine are fixed. After initialization of an *empty_bottleneck_list*, , the lower-bound makespan (*$makespan_{LB}$*) is calculated as the longest path in the graph representation of the partial solution. Then, the pseudo due dates (*due_ dates*) are calculated as in the SBH. In the subsequent iteration over all machines, the minimal tardiness *$tardiness_{min}$* with respect to the pseudo due dates is calculated for each machine and added to the lower-bound makespan to form the bottleneck information (*bottleneck_info*) of that machine. More critical machines are characterized by a higher bottleneck information value. After min-max scaling the vector of bottleneck information of all machines, the final bottleneck information vector (*bottleneck_info_list*) is obtained.

4.2.2 Base Models and Bottleneck Feature Integration

The study involves training and testing two base models that serve as learned construction heuristics trained with DRL. One base model is a straightforward self-designed model architecture that relies on engineered features and a comparatively small NN architecture. In its base design, it smoothly converges towards a policy that is superior to PDRs but leaves much room for potential improvement. The second base model is one of the best-performing DRL-based designs for the JSSP in the literature [74] and uses both, raw data and engineered features. It was chosen because it involves a larger NN architecture that may be more capable of using the bottleneck feature to a positive effect, however, it leaves less room for improving the final policy. In both cases, the training and final performance are compared with and without the additional bottleneck feature. To ensure that potential differences can be attributed to the quality of the bottleneck feature, the bottleneck feature vector is set to 0 whenever the bottleneck feature is not used. This way, the number of parameters in the NNs remains the same with and without the bottleneck feature. All experiments are conducted on JSSPs of size 6×6 and 10×10 and repeated with three random seeds.

Self-designed Base Model

The goal of the self-designed base model is not to outperform the existing state of the art but to provide a stable baseline with good convergence behavior to facilitate the detection of differences to the learning behavior and final performance between those training runs with and without the bottleneck features. The design choices, features and hyperparameters were found in preliminary experiments to satisfy this goal sufficiently and leading to makespans that outperform the common PDRs MTR, SPT and RANDOM. All experiments are conducted in the software framework *schlably* [223] that was created as part of this study. PPO with objective clipping [52] and action masking [205], to exclude already completed jobs form the action space, is used as DRL algorithm. The separate policy and value networks are multilayer perceptron networks with two hidden layers, each with 256 neurons and Tanh activation function. The observation vector contains concatenated information on the next unscheduled operations per job, described by seven features:

1. The processing time of the operation
2. The remaining processing time of the corresponding job
3. The number of remaining operations of the corresponding job
4. The remaining processing time on the required machine
5. The number of remaining operations on the required machine
6. The earliest starting time of the operation given the current planning status
7. The MTR-feature

Since the agent decides between the next operations per job, the collection of features is designed in an operation-centric way, describing job related (1–3, 6, 7) and machine related (4, 5) features. Furthermore, the features include short-sighed features about the next operation (1, 6) and far-sighted features about representations of the sum of remaining work (2–5, 7). Furthermore, all features correspond to information needed to calculate the job prioritizations through common PDRs, such as SPT (feature 1) or MWKR (feature 2). Most explicitly, this is done in the

MTR-feature, because it is the best-performing PDR on average. It is a vector of the pair-wise comparison between the number of remaining operations for all jobs, where a 1, −1, and 0 indicate if the job has a larger, smaller, or equal number of operations remaining. Each feature is min-max scaled across jobs. Noteworthily, in most cases and even before min-max scaling, the feature vector does not uniquely describe the problem instance, meaning that there are several problem instances which lead to the same initial observation.

Whereas the seven own features are calculated per job, the bottleneck information of the SBH is calculated per machine. For the logical integration of the bottleneck information we proceed with the logic used in machine-related features 4 and 5, using the bottleneck information of the required machine per job. Notably, this means that some bottleneck information may not be included in a step, if a machine is not required by any of the next operations. The vector is concatenated to the original feature vector. As stated earlier, the bottleneck information is replaced by 0's in comparative experiments to maintain the number of parameters in the models with and without the bottleneck information. The integration of the bottleneck information and general information flow is depicted in Fig. 4.2.

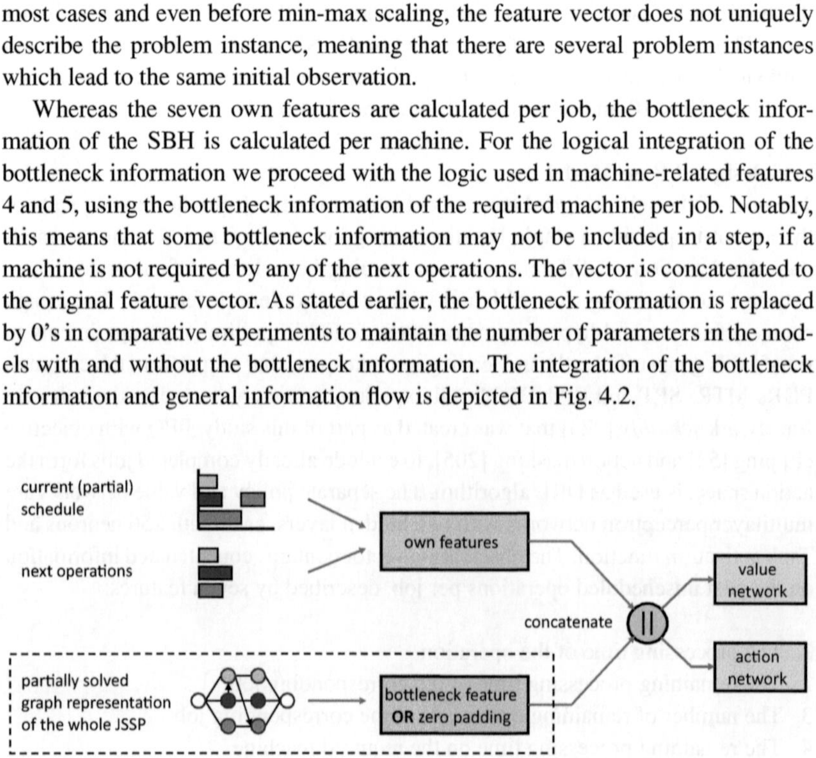

Fig. 4.2 Bottleneck feature integration and data flow in the self-designed base model

The training, validation, and testing datasets and all procedures are adapted from the *schlably* [223] framework with the hyperparameters summarized in Table 4.1. Hence, the training instances are generated through the random assignment of machines to operations within a job and with operation times drawn from a uniform distribution of integers in the interval [1, 11]. The agents are trained on a fixed set of 1,200 training instances per problem size for 4,000,000 steps (6 × 6 JSSPs) and 700,000 (10 × 10 JSSPs), as no further improvement was observed beyond this point in preliminary experiments. The model is validated on a separate validation data set of 40 instances every 30,000 steps and a copy of it saved, if the achieved mean makespan is below the one of all previous models. The final test is performed on another separate set of 40 test instances.

Table 4.1 Hyperparameters of the self-designed base model

Parameter	Value
Algorithm	PPO-masked
Clip-Range	0.2
Batch size	256
Ent_coef	0
Gamma	1
Learning rate	0.02
N_epochs	5
Policy_activation	Tanh
Policy_layer	[256, 256]
Value_activation	Tanh
Value_layer	[256, 256]
Rollout_scale	2048

Learning-to-Generalize (l2g) Base Model

The l2g base model is an adaption of an approach by Iklassov et al. [74] which achieves state-of-the-art results among DRL-based learned construction heuristics. The main contributions of the corresponding publication are the introduction of an expressive, size agnostic NN architecture for the JSSP and the introduction of CL strategies that leverage the size agnostic network design by purposefully changing the size of training problem instances over the course of training. For the means of the bottleneck information study, the reportedly best performing CL strategy and hyperparameters are used. In this way, complimentary to the self-designed base model, it can be tested whether a more sophisticated NN architecture can leverage the bottleneck information effectively even with comparatively little room to improve the optimality gap.

For the bottleneck information integration, two modifications are made: The first modification lies in the problem sizes that are included in training. A separate model is trained per problem size without a curriculum and on instances of smaller problem sizes (6×6 and 10×10), to reduce the computationally expensive bottleneck information calculations in the experiments. A drop in absolute achievable optimality gap on benchmark datasets, which is likely given the substantial changes to the curriculum learning strategy, is accepted, since the purpose of the study is the relative difference with and without the bottleneck information. The second modification to the originally proposed method is caused by adding input features into the NN. As depicted in Fig. 4.3, the original l2d architecture creates separate embeddings of the static representation of the problem instance and a dynamic representation consisting of the three feature vectors: last processed operations per job, the machine status, and the remaining processing times. These embeddings are combined at later stages of the network. The bottleneck information is added as a fourth dynamic feature vector. To this end, the bottleneck features per machine required for the next operation are concatenated to the existing dynamic feature vector and then projected to the embedding size 128 through a linear layer. For a detailed description of the other aspects of the network architecture, the reader is referred to the publication by Iklassov et al. [74].

Otherwise adhering closely to the original training procedure, all training and testing instances consist of operations with processing times sampled uniformly from the interval [1, 100]. The instances are created separately from those of the study on the self-designed base model. Every episode, new training instances are randomly created. During training, the current model is evaluated on ten fixed evaluation instances every 100 episodes and its parameters are saved, if the achieved average makespan is smaller than the best observed average makespan up to that point. The resulting best performing model is finally tested on ten separate test instances. Training is stopped after 2,800 episodes and 1,000 episodes for the 6×6

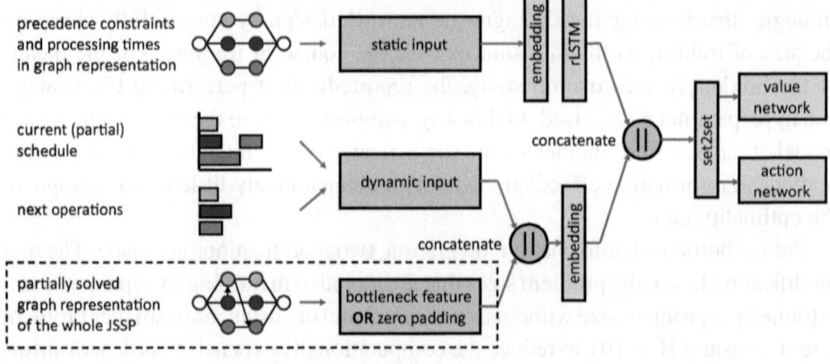

Fig. 4.3 Bottleneck feature integration and data flow in the l2g base model

and 10 × 10 JSSPs, respectively, as no further improvement was observed beyond that point with or without the inclusion of the bottleneck information in preliminary experiments.

4.3 Results

The results of the training processes and model tests are aimed at two related yet separate hypotheses: the first hypothesis is that providing the bottleneck information directly is more effective than relying on the agent to infer bottleneck information from the rest of the input, such that the learning speed should increase. The second hypothesis is that the bottleneck information is so useful that its inclusion provides an advantage to the agent that it cannot compensate with processing raw information of the problem instance and information on the current schedule, such that the inclusion of the bottleneck information leads to a better final performance. These hypotheses are analyzed in the following section.

For the visual comparison of the convergence rate of the DRL agents, the achieved average evaluation makespans on the respective problem sizes during training are plotted over the number of instances seen during training. The corresponding plots are shown in Fig. 4.4 with solid lines representing the average values across three seeded training runs and the shaded areas indicate the standard deviations thereof. The results of the model trained with the bottleneck feature are plotted in blue and the ones without in orange. Generally, the evaluation results drop fast during the first couple of hundred training instances and then converge towards a final plateau around which the models oscillate. Critically, no clear differences appear between the two base models, neither in convergence speed, nor in the evaluation makespan, nor in the oscillation behavior. From the visual inspection it therefore seems that the inclusion of the bottleneck information does not change the learning behavior, falsifying the hypothesis about a positive impact.

The second part of the analysis evaluates the final performance the trained models achieve on the test dataset. The averaged results across the three seeded training runs are reported in the *agent* column of Table 4.2. The column *optimal* reports the average makespans of optimal solutions that were calculated for the test data set using the CP-SAT solver of Google's OR Tools library [34]. The *opt.-gap.* column is the percentual gap between the optimal solution and the average agent solution:

$$opt. - gap = \frac{agent - optimal}{optimal}$$

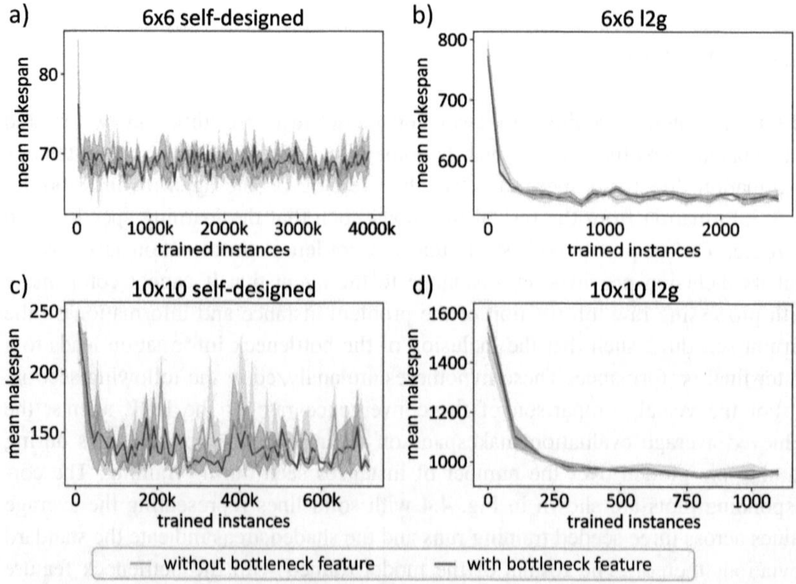

Fig. 4.4 Learning curves of the self-designed and l2g base models

The standard deviation of the achieved average results and optimality gaps are indicated behind the plus-minus signs. For comparison, the deterministic results of *MTR*, *SPT* and *RANDOM* prioritization rules are printed in the last three columns. In comparison to these rules, it is evident that both base models outperform all prioritization rules, except for the self-designed model on the 10×10 JSSPs performs worse than MTR. The critical results with respect to the working hypothesis are the direct comparison between the same base models on the same problem sizes but with and without the bottleneck feature. Here, confirming the pattern observed in the learning curves, no significant distinction between the test results can be observed with most differences in achieved average makespans smaller than 1.4 %p, and all differences lying within the standard deviations of the three seeded runs. Furthermore, there are two cases in which the models with

Table 4.2 Summary of average test results of the trained models

base model	JSSP size	bottleneck-feature	agent	optimal	opt.-gap	MTR	SPT	RANDOM
self-designed	6 × 6	yes	69.5 ± 2.1	59	17.8 % ± 3.6 %p	72.1	140.7	98.9
		no	**68.7 ± 2.0**		**16.4 % ± 3.4 %p**			
	10 × 10	yes	**123.7 ± 5.1**	96.9	**27.5 % ± 5.2 %p**	117.2	320.1	165.3
		no	123.9 ± 2.5		27.8 % ± 2.9 %p			
l2g based	6 × 6	yes	543.7 ± 1.4	494.8	9.9 % ± 0.2 %p	602.2	1209.6	828.7
		no	**542.0 ± 3.5**		**9.5 % ± 0.7 %p**			

(continued)

Table 4.2 (continued)

base model	JSSP size	bottleneck-feature	agent	optimal	opt.-gap	MTR	SPT	RANDOM
	10 × 10	yes	**962.2 ± 3.9**	827.7	**16.3 % ± 0.4 %p**	1094.6	2910.0	1599.3
		no	963 ± 2.5		16.4 % ± 0.5 %p			

the bottleneck information perform slightly better (10×10 JSSPs) and two where they perform slightly worse (6×6 JSSPs).

Considering these results, the additional non-zero features containing information of the bottleneck feature inspired by the SBH seem to introduce noise to the experience data rather than to provide useful information for the learning task. It should also be noted that both, the inclusion of the bottleneck feature and that of zeros, had a small detrimental effect on the final performance compared to preliminary experiments without the additional feature vectors but with the same training procedures. This shows that features containing little or no informative value confuse the model rather than increase the learning speed or final performance. Although this behavior has been previously mentioned in other studies on DRL-based scheduling algorithms [40, 224], it has never been shown for the JSSP.

4.4 Summary and Discussion

The study in this chapter presents a comparative analysis of the training speed and final policy of DRL agents with and without a bottleneck feature inspired by the SBH. Experiments were conducted on a self-designed DRL algorithm and a state-of-the-art DRL algorithm on the 6×6 and 10×10 JSSP. No significant difference was observed between those training runs with and without the bottleneck feature, leading to the conclusion that the proposed bottleneck feature, in its current form and integration, does not provide any benefit to DRL-based JSS algorithms.

Nevertheless, the study provides insights into aspects of RQ1, which asks *how domain knowledge can be advantageously integrated into the observation space.* This chapter demonstrates that bottlenecks are an integral part of many problem descriptions and designed solutions in the JSS literature. This makes them a well-motivated candidate for domain knowledge that can be integrated into the observation space of DRL algorithms. It was shown that bottleneck information can be integrated into the observation space in the form of supplementary machine-wise features. However, the results indicate that, at least in the specific configuration of the SBH-inspired bottleneck feature in conjunction with the two selected base models, the additional information yields no advantage with respect to training speed or the performance of the final policy. Consequently, the demand for an "advantageous integration" in RQ1 remains unanswered.

One notable limitation of the study is the selection of the bottleneck feature. The choice of the SBH-inspired feature was primarily motivated by the success of the heuristic and the distinct differentiation from the typical machine features observed in the existing literature, which greatly enhances the novelty of the presented method. Nevertheless, in retrospect the utilization of the SBH bottleneck definition may have been an unfavorable choice for two main reasons: The first reason is the discrepancy in the iteration logic between the SBH and construction heuristics. In the SBH, a bottleneck machine is identified and subsequently, all operations on that machine are ordered within fixed constraints. However, this order may be altered again in later iterations. In contrast, the DRL-based construction heuristics fix exactly one operation permanently in each iteration. The difference in this logic not only requires deviations from the way the bottleneck information is originally computed. It also weakens the intuition that the bottleneck definition, despite empirically working well in one iteration logic, translates well into another logic. The second reason why the selection of the SBH bottleneck information may have been suboptimal is the substantial computation time required for its computation during both training and inference. The incorporation of the feature results in approximately 16-fold longer computation times for the 6×6 JSSP, thereby mitigating the inference time advantage DRL-based methods have over exact solvers and metaheuristics. The primary factor contributing to the increase of computation time is solving the NP-hard one-machine problem for each machine in each iteration. For this reason, training and testing on larger JSSP sizes would have taken prohibitively long. A potential remedy for the long computation time would have been the approximation of the bottleneck information calculations through another deep learning model. However, this possibility was not pursued after the non-approximated feature proved ineffective.

Furthermore, certain methodological limitations of the conducted study must be considered in light of the broad scope of RQ1 and the comparatively narrow results. One limitation of the study lies in the limited scope of the analyzed candidate bottleneck information features. The definition of bottlenecks employed in the SBH is only one possibility among many metrics of bottlenecks, and others may have had a positive impact. Broadening the scope even further, bottlenecks are only one of many possible elements in the acquired body of domain knowledge on the JSSP in the literature from which DRL-based methods could profit. This leaves much room for future work in this area.

Nevertheless, the explicit integration of bottleneck information remains a promising approach for the improvement of DRL-based methods. In future work, the efficacy of other measures for the bottleneck that are aligned with the construction heuristic approach could be examined, such as measures created from model-based lookahead simulations. In such experiments, it might prove beneficial to adapt modern, more expressive deep learning architectures that facilitate the integration of bottleneck information, such as the self-attention decoder-based architecture suggested in [57].

Taking a step back, there are other means of integrating bottleneck information into the DRL algorithm apart from features in the observation space. For example, an efficient utilization or prioritization of bottleneck machines could be incentivized through the reward design. Alternatively, a modification to the action logic from a conventional construction heuristic to a logic in which operations may be iteratively re-ordered on a bottleneck machine, like in the SBH, could yield better results. It is important to note that these concepts diverge further from the research question posed in this chapter.

Lastly, the notion of bottlenecks represents only a small part of the domain knowledge pertaining to the JSSP. Other problem characteristics, such as the solution symmetry, are still underrepresented in the DRL-based JSS research. Hence, they remain another promising direction for future research.

5 Effective Learning Curricula for the JSSP

It can be posited that DRL, in comparison to supervised and unsupervised learning, is the machine learning paradigm that most closely mirrors the way in which humans learn—by interacting with the environment, experiencing positive or negative feedback, and changing the behavior in response. However, in DRL-based JSS, certain aspects that contribute to the success of human learning are modified and studied more frequently than others. The review of the existing literature on DRL-based JSS, as introduced in Chap. 3, reveals that most studies introduce at least one innovation along the following dimensions: the design decision for the interaction logic between the agent and the environment, the reward formulation, the feature design for the observation, the DRL algorithm, and the NN design. While some contributions for advanced training regimes are gaining traction, rather little attention is devoted to the generation and distribution of training data over the course of the training process. At a minimum, such design choices are seldom explicitly analyzed with respect to their influence on the achieved results. This stands in stark contrast to the development of learning curricula for teaching human children, where significant emphasis is placed on the question, how topics and skills build on top of each other and what a suitable sequence of these topics and skills would be for educational implementation in schools. The artifacts of such considerations in human learning are mandatory schooling curricula. By exploring the potential for advancing learning regimes for DRL-based JSS analogously to established methods for human learning, this chapter aims at shedding light on RQ2, *how training strategies should be designed for efficient and effective DRL agent training* on the JSSP.

In the machine learning community, the concept of curricula has already been successfully adopted. The term "curriculum learning" (CL) was originally coined

C. Waubert de Puiseau, *Advancing Training and Inference Methods for Reinforcement Learning-Based Job Shop Scheduling*,
https://doi.org/10.1007/978-3-658-51305-4_5

by Bengio et al. in 2009 [225]. The concept builds upon the notion that starting training on easy tasks and gradually increasing the difficulty, much as in human learning, is beneficial for training NNs. Since its introduction, CL has evolved into a distinct and active research field [8, 226, 227], in which definitions and methods have become more nuanced over time. In the broadest sense, CL involves the targeted modification in the training setup over the course of training to enhance convergence speed or improve the final performance of a machine learning model [226].

For supervised learning, where most of the early development took place, Soviany et al. [226] distinguish between data-level and model-level CL. In the former, the training data is provided in a deliberate sequence during the training process, whereas in the latter, the architecture of the model is changed over the course of training. Most methods belong to the class of data-level CL algorithms [226]. In their survey, Wang et al. [8] identify two critical components that are involved in data-level CL methods: the *difficulty measurer* and the *training scheduler*. The difficulty measurer assigns a difficulty score to each datapoint in a given dataset, laying the foundation for sorting the data accordingly. Common difficulty scores can include, but are not limited to, the number of classes in a labeled dataset or the episode length in a RL setting. The training scheduler determines when to utilize a specific batch of the sorted training data during training. Such a schedule can be predefined before training starts or adapt during training based on, for example, an achieved loss threshold on a training data batch or even by a meta-DRL agent trained for this purpose. [8]

In recent years, CL has also been applied to DRL. In this context, one of the latest surveys by Narvekar et al. [227] proposes a different framework for the categorization of CL methods. This framework encompasses three elements: *task generation*, *sequencing* and *transfer learning* [227]. Task generation is defined as "the process of creating a good set of intermediate tasks from which to obtain experience samples" [227]. Sequencing determines the method dictating or adjusting the sequence in which tasks are introduced over the course of training [227]. The concepts of task generation and sequencing are closely related to the difficulty measurer and training scheduler introduced by Wang et al. for supervised learning [8]. The primary difference lies in the emphasis placed on an explicit transfer learning method. Transfer learning is defined as the study of learning in a source domain and applying useful knowledge gained there in a different target domain [228]. To illustrate, knowledge gained from learning to discriminate between cat and dog images can be applied to discriminate between different dog breeds. Transfer learning is particularly relevant in the context of CL, as shifts in the distribution of tasks over training are analogous to domain shifts.

The efficacy of CL in terms of accelerating training or attaining better final policies has been empirically substantiated in various DRL settings. Different intuitions have been postulated to explain why CL is so effective in these cases. One such intuition, as depicted in Fig. 5.1 and articulated by Bengio [229], is that the initial task of a successful CL strategy has a smoother objective function (depicted in green) and it is thus easier to find a global minimum via stochastic gradient decent along this objective. Subsequently, harder but related tasks will have their global minimum in proximity to that of the first, easier task. This explanation is very intuitive for the cats-vs.-dogs classification and subsequent dogbreeds classification curriculum example that was mentioned above. However, the intuition may not hold when the found local minima of subtasks lie far apart. For example, if a model has learned to discriminate dogs and cats by identifying cat whiskers, this strategy will not transfer well to the dog-breed classification task.

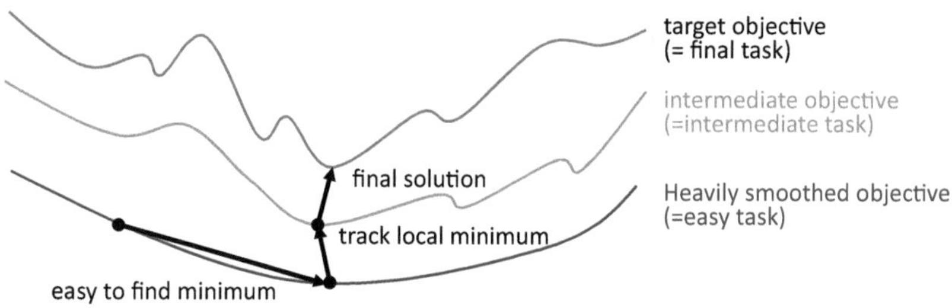

Fig. 5.1 Intuition that easy tasks have an easy-to-find minimum in CL (adapted from [230])

Another intuition for CL is that an agent, much like a human, could first learn a fundamental concept through solving an elementary task and then later apply the concept in a more difficult task. An example for this intuition in human learning would be learning the concept of mathematical division for single digit integers before applying division to floating point numbers. However, not every task is separable into elementary sub-tasks. A third intuitive perspective is that it is easier to learn fundamental tasks separately and later combine them. An illustration for this intuition would be a robotic pick-and-place task, for which it could be beneficial to first learn how to pick something and separately how to place something, before combining the two learned skills. Evidently, the intuitiveness of these explanations varies in different problem settings. However, in certain problem settings in which subtasks and difficulties are not as straightforward to define, such as in DRL-based JSS, the explanations for how and why CL could be successful are less intuitive.

Yet, in 2023, Iklassov et al. [74] published a study on a DRL-based method for the JSSP. This study introduced a novel DL architecture for the JSSP and a CL strategy for training this architecture, reporting that using their CL strategy leads to around 6 %p—12 %p smaller optimality gaps than training without a CL strategy. Further details on the study are given later in SubSection 5.1. The sensitivity to CL as well as the substantial induced improvement in the final policy through variations in the CL strategy motivates the further study of CL methods for the JSSP which is conducted in this chapter.

Accordingly, the remainder of this subchapter is structured as follows: SubSection 5.1 provides an overview of related work on CL for JSS. In SubSection 5.2, the new CL strategy is motivated and developed, and the experiment design to test the strategy is described. Subsequently, the results of the study are presented in SubSection 5.3. Finally, the results are discussed with respect to the contribution and limitations regarding RQ2.

5.1 Related Work: Curriculum Learning for JSS

Several studies have examined the role of CL in solving combinatorial optimization problems through NCO methods. In 2020, Lisicki et al. [230] asserted that many NCO models are adept at learning on and generalizing to multiple problem instance sizes. Consequently, they investigated the literature on CL as a remedy. In their experiments, the authors demonstrated that training on a deliberate curriculum is superior to training on randomly arranged training datasets. However, the experiments also showed that models experience catastrophic forgetting when switching between tasks during training. However, rehearsal methods, defined as periodically returning to previous tasks, provide remedy and allow for effective training on multiple problem instance sizes. The work by Lisicki et al. [230] differs from the work presented in this chapter, as it focuses on the TSP and employs CL for a supervised learning method, whereas this chapter deals with the JSSP and RL. Concurrently, in 2020, Nabil et al. [231] studied the application of CL in multi-agent DRL to solve several multilevel budgeted combinatorial optimization problems. Specifically, their method increases the problem size of training instances once a performance threshold is reached on the current problem size. The authors of this study also conclude that this CL method is superior to using no curriculum. However, in contrast to the work presented in this chapter, no scheduling problems are examined.

Zhou et al. [232] took a formal approach to studying CL in combination with the vanilla policy gradient RL method (cf. SubSection 2.2). Their definition of tasks in a CL strategy is restricted to certain lengths of online combinatorial optimization problems where only acceptance or rejection decisions are required in each decision step. For this method and problem setup, the authors provide a mathematical proof that any curriculum improves the convergence speed exponentially over using no curriculum, i.e., training on random problem lengths. This proof is also experimentally verified. However, one cannot conclude that the results are necessarily transferable to more advanced DRL algorithms that involve value approximation networks, or to other combinatorial problems.

All three studies point in the same direction that CL is a valuable method for increasing the efficiency and final performance of DL models applied to combinatorial problems. However, none of the studies explicitly involves the JSSP or evaluates curricula that are not solely based on the problem size. Therefore, the transferability of insights is not guaranteed, especially in light of JSSP-specific characteristics such as the large solution symmetry, which are not as pronounced in the addressed combinatorial problems (cf. Sect. 2.1.2).

Nevertheless, the notion of the potential of CL methods for scheduling problems has also begun to manifest itself in the scheduling literature. Liu et al. [120] designed and evaluated a multi-agent DRL approach for the dynamic JSSP. They observed that agents that were trained with different dynamic settings of the environment built different generalization abilities when tested in less dynamic settings. This led the authors to suggest that CL techniques exploiting this phenomenon should be investigated in future work [120]. CL was also identified as an important future trend for DRL-based JSS to improve generalization in a literature review by Zhang et al. [233] in 2024. However, both studies do not share any explicit insights on how to design a suitable CL strategy.

Certain training designs in the literature steer the learning behavior of DRL agents by changing the reward signals that are increasingly difficult to receive, thus providing an implicit training curriculum from easy to difficult on the same problem. Samsonov et al. [58] introduce a novel multi-instance ranked reward method that achieves this effect. The method maintains a buffer of previously achieved rewards on the training dataset against which the results of the current training episode are compared. The agent receives a positive reward, if the average makespan achieved in the current epoch is smaller than that of the previous epoch, a negative reward if it is greater, and positively or negatively with a certain probability to incentivize exploration. Pirnay et al. [78] take a slightly different approach. Here, an agent competes against other and earlier versions of itself, which provides a performance baseline on the JSSP. If the agent beats a certain threshold makespan compared to the baselines, the sparse reward at the end of an episode is positive, otherwise it is negative. Both studies show that the reward structure is superior to static reward designs. Although it can be argued that the difficulty of tasks is implicitly varied in these reward designs and that they are therefore related to CL, the overall input instances and the overarching goal of minimizing the makespan remain the same.

Only few studies explicitly introduce CL for machine scheduling problems. Hameed et al. [234] trained a DRL agent that can handle a varying numbers of jobs in the system. They start training with a limited number of jobs but a fixed number of machines and gradually increase the number of jobs over the course of training, following the intuition that this strategy gradually increases the difficulty for the agent. Unfortunately, they do not provide an ablation study with respect to the curriculum, so an analysis of the contribution of the CL strategy to the results cannot be determined. However, it is likely that they saw a benefit of CL in preliminary trials. More explicitly, Iklassov et al. [74] compared several different CL strategies against each other and reported significant differences in the final achieved performance of the respective agents. As done in the aforementioned studies, different learning tasks are discriminated by problem size. Specifically, Iklassov et al. [74] use the problem sizes 6×6, 10×10, 15×15, 20×15, 20×20, 30×15 and 30×20, in that order, as increasingly difficult problems. They define three curricula and an incremental baseline: Incremental Curriculum Learning (ICL) trains a separate model for each problem size. Adaptive Staircase Curriculum Learning (ASCL), adapted from [230], starts with the easiest task and repeatedly evaluates the current agent on the current problem after a defined number of training steps. If the agent performs above a performance threshold, it is subsequently trained on a task that is one step more difficult. If it falls below a performance threshold on the current task, it is subsequently trained on a one step easier task. Their newly developed CL strategy, Adversarial Curriculum Learning (RASCL), introduces an additional switch between tasks after a predefined number of task switches that happen according to ASCL. To determine which task to switch to, the current agent is tested on an evaluation set of every problem size, and the percentual gap to optimal solutions is translated into probabilities of choosing each respective problem size as the next task. The figure summarizing the results of the comparative study are reprinted in Fig. 5.2.

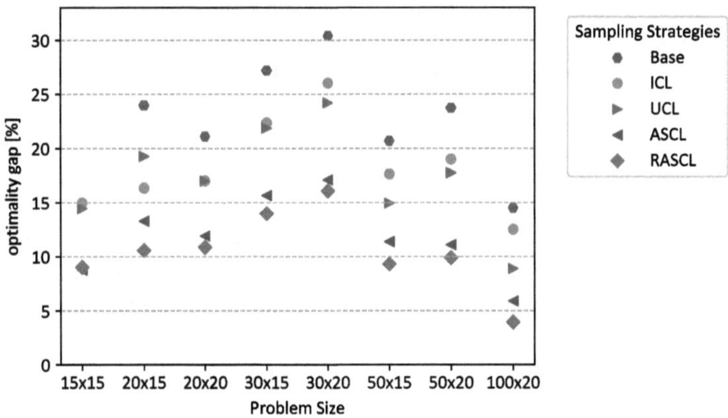

Fig. 5.2 Results of different CL strategies achieved by Iklassov et al. (adapted from [75])

As can be seen in Fig. 5.2, the makespan achieved by the agents varies significantly and RASCL outperforms the other methods on average, leading to the best overall results for DRL base construction heuristics at the time of publishing the results in 2023 [74].

Noteworthily, all aforementioned studies that design deliberate CL strategies define the problem difficulty of combinatorial optimization problems according to the problem size. This is a natural choice, since the formal computational complexity also scales exponentially with the problem size (cf. Subsection 2.1.2). However, it remains an open question, whether a difficulty difference can be defined within a fixed problem size and if such a measure that can be used to form curricula that improve training for DRL-based machine scheduling agents. Scanning the literature reviewed in Chap. 3 for CL strategies, only one study presents an explicit CL strategy with this property. Li et al. [120] define a two-stage CL strategy to minimize the total tardiness of jobs in a machine scheduling problem setting. They claim that minimizing the tardiness is a more difficult DRL task than minimizing the makespan, because the total tardiness is a sparse metric and therefore a sparse reward signal must be used, compared to the makespan, which can be easily represented by a dense reward of locally makespan differences. Yet, they argue that minimizing the makespan is a useful preliminary task, because denser schedules are likely to yield smaller total tardiness values. Thus, in the first stage of the curriculum, the DRL agent is trained with a dense makespan-related reward, before the reward changes to a sparse tardiness-related reward in the second stage. Li et al. [184] confirm their hypothesis experimentally, showing that agents trained with their curriculum learn faster and achieve a better final performance than agents trained directly on the tardiness-related reward. Their work serves as another example that CL is useful for machine scheduling problems, but again, it does not provide a perspective on designing a CL strategy for training a DRL agent to minimize the makespan for a fixed problem size.

In summary, the reviewed related work showcases promising examples for significantly improving the training of ML models for solving combinatorial problems, speeding up training, and leading to better performing trained models. However, a method to leverage CL for the JSSP of a fixed-size to minimize the makespan, as is done in this chapter, has not been investigated.

5.2 Experiment Design

The experimental setup is designed within the framework of the necessary elements for CL in DRL as defined by Narvekar et al. [227], namely task generation, sequencing, and transfer learning. The combination of all three elements constitutes a CL strategy, and changing one of the elements is likely to affect the overall learning behavior. To reduce the number of variables in the experiment design, the transfer learning element is deliberately omitted. There are three main arguments, that motivate omitting this element instead of the others. First, transfer learning is most important when a curriculum consists of clearly distinct subtasks. For learned construction heuristics on the JSSP, there is no natural separation into distinct subtasks, which reduces the need for an elaborate transfer learning method. Second, one can partially compensate for the effect of an explicit transfer learning method by making other design choices that favor strong generalization. For this purpose, a state-of-the-art approach by Zhang et al. [67] is chosen as the base model for the experiments, because it features a GNN architecture, that is known for its good generalization behavior in general [235], but also specifically for the JSSP [19].

Before explaining the choices for the remaining CL design elements, task generation and sequencing, an introduction to the base method developed by Zhang et al. [67] is given to provide the necessary context for understanding the resulting implications for the subsequent design choices for the other two CL elements.

The DRL agent trained by Zhang et al. [67] iteratively chooses the next operation to integrate into the current schedule by assigning probabilities to all unfinished jobs, as is common for learned construction heuristics on the JSSP. However, instead of determining the earliest starting point of the selected operation, the environment performs an additional check. It determines the last operation that is scheduled on the machine required to process the selected operation and checks whether swapping the order of these two operations will reduce the makespan of the intermediate schedule. If this is the case, the order is swapped, otherwise, the original order is kept. The dense reward is equal to the difference between the intermediate makespan before and after the action. The NN architecture, Graph Isomorphism Network [235], makes use of the disjunctive graph representation of the JSSP in which each node corresponds to an operation (cf. Subsection 2.1.2 for a formal introduction). Each node is initially represented by a state vector containing two features, a binary indicator of whether the node has already been scheduled, and an integer lower bound on the estimated completion time of the node. The node vectors are updated by message passing along the graph edges, i.e., between operation nodes of the same job and those sequentially scheduled on a machine. Finally, a global embedding of the current partial schedule is generated by average pooling across all nodes. This last step is crucial, because it results in a fixed-size embedding independent of the problem size. To be fully problem-size agnostic on the output side of the model, the global embedding is separately concatenated with the embedding of each node corresponding to a potential next action, and each resulting vector is processed by a feedforward network to generate an output value per node. Finally, SoftMax is applied to the vector containing all these output values to obtain a distribution of probabilities for actions. As mentioned above, this NN architecture demonstrates very good generalization capabilities, making it a good candidate base model for a CL strategy without explicit transfer learning technique.

Zhang et al. [67] train this network on 40,000 training instances of the same problem size. They train separate networks for small to medium problem sizes, 6×6, 10×15, 15×15, 20×20, and 30×20, of the Taillard benchmark [22], and tested on all problem sizes, including 50×15, 50×20, and 100×20, of the benchmark. Reportedly, training on larger problem instances took too long, and the generalization to other problem sizes allowed the authors to still apply their trained networks to them for inference. The goal of the study of CL in this chapter is to analyze the relative impact of a CL strategy versus using no CL strategy. Therefore, the CL approach, in this case the remaining task generation and sequencing approaches, should be designed to also work with a fixed dataset. This is achieved through the following method.

For the task generation from a dataset of problem instances of the same size, a rule for building clusters or a measure of difficulty is needed. Instances are described by the order in which operations must be scheduled on a machine and by the processing times of their operations. From this information alone, it is non-trivial to derive a formula that is representative of a difficulty measure. Problem instances of the same size have equal theoretical complexities (cf. SubSection 2.1.2), which eliminates the possibility to use them as difficulty measure. It would instead be possible to calculate the standard deviation of all processing times within an instance or compute the difference of the sum of processing times on each machine. Although such measures are possible candidates, they are not grounded in any theoretical or intuitional reasoning as to why instances with a particularly small or large resulting scores would pose a more or less difficult task for an agent. A more intuitive way would be to create a solution of instances and develop a measure that characterizes this solution, such as the ratio of the sum of all processing times to the makespan, which is an indication of the achieved machine utilization. However, this would require solving many instances perfectly, which is very computationally expensive. Yet another perspective would be to define an agent-centric measure by testing how well the agent performs on an instance relative to a heuristic. However, especially in the beginning, policy-based agents change their behavior very quickly, which would require very frequent updates of the difficulty measure of all training instances, rendering such an approach computationally very expensive. There are countless more options, some of which are discussed in SubSection 5.4 and left for future work.

The method developed in this study is incentivized by the perspective of a hypothetical user, for whom a problem instance that can be solved well using a simple prioritization rule appears as "easy to solve" and one that cannot be solved well appears as "difficult to solve". Solving an instance "well" can be defined in relative terms to the optimal solution by measuring the optimality gap of the solution. However this requires computing optimal solutions for the entire training dataset, which, as mentioned above, becomes computationally intractable. Instead, solving an instance "well" is defined here in absolute terms of the makespan and then relative to other problem instances. Specifically, the best-performing priority rule in combination with the scheduling logic in the DRL environment of Zhang et al. [67], the most-tasks-remaining (MTR), is applied to each problem instance in the training dataset and the achieved makespan used as a "difficulty-to-solve" measure (DTS). According to this logic, those problem instances with a DTS smaller than average makespan can be called easy and those with a larger than average makespan can be called hard. For this study on CL, we randomly create 40,000 6×6 JSSP training instances according to the code provided by Zhang et al. [67] with a random seed of 401. The smallest problem size is chosen to keep the required computational time of the experiments short, as will be discussed in more detail in SubSection 5.4. The instances are subsequently solved using the MTR prioritization. The resulting makespans are shown in the histogram, visualizing the number of occurring DTS values, in Fig. 5.3. It can be seen that the distribution of the DTS values resembles a normal distribution, with a calculated standard deviation of 70.3 time units.

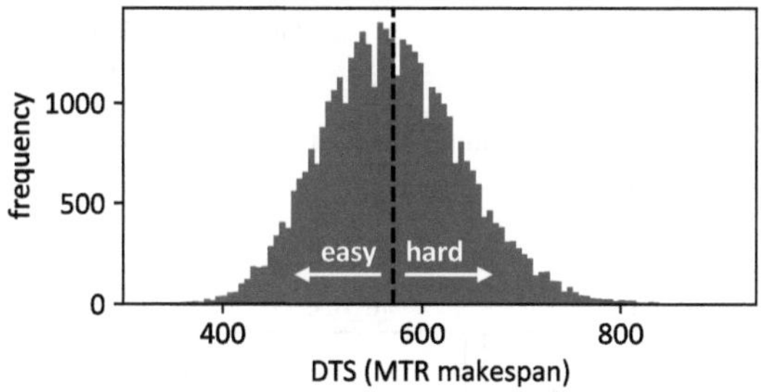

Fig. 5.3 Histogram of the DTS values of 40,000 training instances

The next step is to determine a suitable sequencing technique. There are many ways to build curricula from the nearly continuous DTS measure. As has been shown in the past, a fundamental question in CL is whether one should start with easy or hard training data [227]. A second point of interest, especially for CL in DRL, is the transition from one "task" to another, e.g., the immediate change from training on easy instances to training on hard instances. In contrast to methods with different discrete subtasks, the continuous DTS measure allows to design a smooth transition or to introduce a deliberate step between training on the easiest and the hardest instances. To achieve the latter, one can divide the training data into two sections, easy data (*e*) and hard data (*h*), and to sort each section from easy to hard (referred to as normal sorting, *n*) or to reverse this order (referred to as reversed, *r*). The four combinations result in four building blocks, which are referred to as curricular blocks from here on, of 20,000 training instances each, *e_n*, *e_r*, *h_n* and *h_r*, as shown in Fig. 5.4.

In the next step, each pairwise sequential combination of these curricular blocks is formed, defining 16 preset curricula that are shown in Fig. 5.5. Noteworthily, this results in smooth transitions in some cases, like [*e_n*, *e_r*] or [*h_r*, *e_r*], but also in non-smooth transitions, most drastically in [*e_r*, *h_r*] and [*h_n*, *e_n*]. It should further be noted that in half of the cases the curricula contain either only the easy or only the hard half of all training instances.

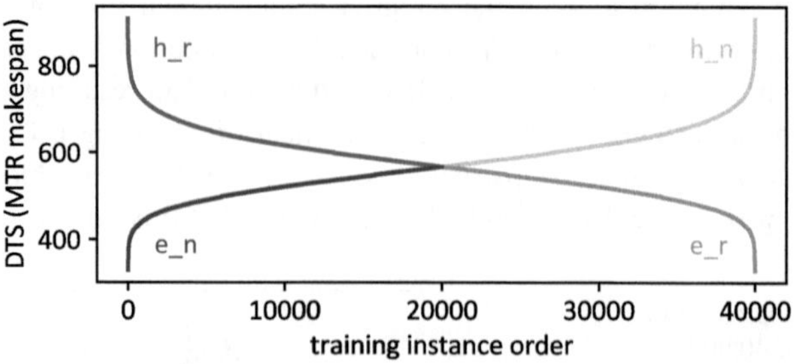

Fig. 5.4 The four building blocks derived from sorting all 40,000 training instances by their DTS measure

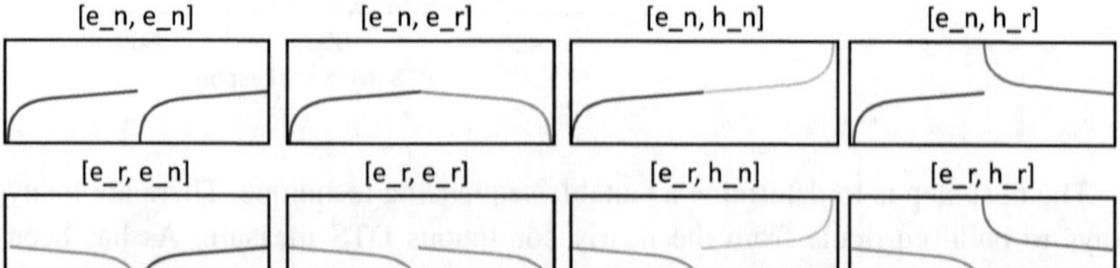

Fig. 5.5 Curricula depicting the sequences of training instances sorted by DTS measure

In summary, the designed CL experiments build on the existing code base of Zhang et al. [67], including all hyperparameters. As a novelty, the training tasks are defined according to blocks of training data sorted by their DTS value measured by the makespan obtained by applying the MTR priority rule, and these blocks are combined into 16 distinct curricula. No specific transfer learning method is applied.

The goal of the experiments is to analyze the impact of the different curricula. To this end, a separate RL agent is trained on 40,000 instances arranged in the

16 curricula depicted in Fig. 5.5. As a baseline, we also train an RL agent on unordered training data, where the problem instances in the dataset are randomly shuffled. The baseline is exactly the way in which the agents were trained according to the original publication [67]. Each training run, the 16 CL runs and the baseline run, is repeated three times with different random seeds for more statistically significant results. This way, the agents are initialized differently in each run and the sampled 40,000 instances in the training dataset are also different. During training, every 2,000 training instances have been shown, the intermediate agents are evaluated on a separate evaluation dataset containing 1,000 problem instances.

5.3 Results

The evaluation results over the course of training are plotted in Fig. 5.6. The y-axis represents the optimality gap in percent, while the x-axis represents the training progress on the scale of how many training instances have been used up to each point. Note that the color scheme in the legend refers to the curriculum building block used during the first half of training. Therefore, the colored solid lines show the average value across four curricula, each with three seeded runs, amounting to a total of 12 runs per colored solid line. The shaded areas represent the minimum and maximum values achieved in the twelve respective runs. Similarly, the black lines indicate the three seeded baseline runs with the randomly perturbed dataset. The first values are recorded after the first evaluation is performed after 2,000 training instances have been shown. Generally, one can observe that the evaluation optimality gap decreases (improves) rapidly in all runs and finds a local minimum after 6,000 training instances, before continued training leads to a worse performance at first, and then, in most cases, to a gradual convergence towards a final optimality gap.

The point after 6,000 training instances have been used for training is an interesting intermediate step, as more than 70 % of all agents reach their global minimum at this point. This indicates that the agents develop the most successful strategy at the very beginning of training, then diverge and never return to it. Furthermore, the top 10 % of all agents in terms of the globally achieved optimality gap all reach their minimum evaluation scores after 6,000 training instances. It is therefore interesting to take a closer look at the zoom-in that is provided on the right side of Fig. 5.6. It reveals that the lowest dip of most averaged runs is directly related to the easiness of the initial data shown to the agent, and thus directly related to the training curriculum. The lowest point corresponds to agents trained with *h_r* as the first curricular block (blue line), meaning that the agent has been trained on the hardest available instances in the training dataset up to that point. Conversely, the red line reaches the highest average value of all colored solid lines after 6,000 training instances. These agents have been trained on the easiest training instances, *e_n*, up to this point. The other two lines complete this trend.

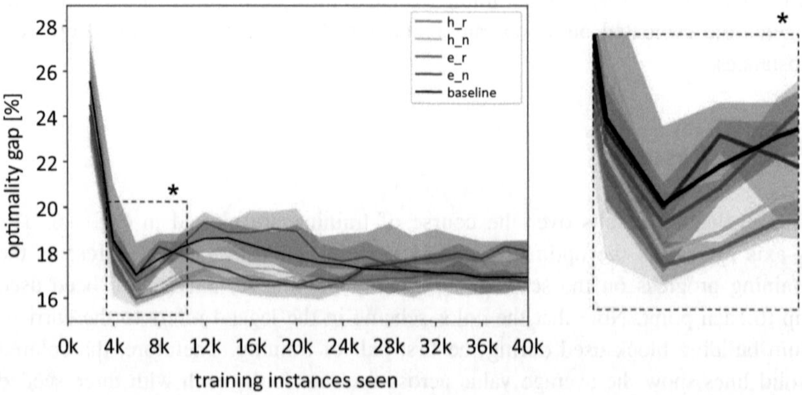

Fig. 5.6 Evaluation scores across training according to different curricula

Noteworthily, all agents trained with a specific curriculum perform better than those trained without a curriculum, as indicated by the dip at 6,000 of the black line. From these observations, one can conclude that it is advisable to use a curriculum, specifically the *h_r* curricular block, at the beginning of the training, to increase the chance of achieving better results. In the presented case, the agents trained with *h_r* at the beginning achieve on average 7 % better results (1.1 %p) after 6,000 training instances. The average across all curricula is about 3 % (0.5 %p) better than the performance of the baseline runs.

Although, as argued above, the first local minimum is the global minimum for most agents and for all best agents, it is not the only point of interest regarding the effect of CL. The second point of interest is after the first 20,000 training instances have been shown, and the second curricular block is introduced. Two main questions are of interest: first, whether the second curricular block has a consistent impact on the final result to which learning converges in the second half of training. If this is the case, such an insight would be valuable for steering the learning process of DRL agents that do not experience the first dip observed in this study. Second, whether the second curricular block has a consistent local effect shortly after its introduction, i.e., on the difference between the evaluation results after 20,000 and 22,000 training instances. If so, the introduction of a new curricular block could be a viable option to escape from local minima during training.

To analyze the impact of the second curricular block on agent training, the evaluation scores after 20,000 training instances are plotted in Fig. 5.7. In this figure, the runs are grouped by the first curricular block on which they were trained, into *h_r* (top left), *h_n* (top right), *e_r* (bottom left), and *e_n* (bottom right). Note that for this reason, their evaluation curves overlap until 20,000 training instances have been shown and the second curricular block starts. The plots are colored according to the second curricular block on which the agents were trained, and the lines are averaged values over three seeded runs. The shaded areas for the minimal and maximal values from the previous figure are omitted for clarity, as they provide no additional insight here. For reference, the miniature full evaluation curves are inserted in each plot. There are only few consistent patterns regarding the curricular block used in the second half and the training behavior of the agents. One could have expected that *e_r* (green) or *h_n* (orange) provoke more changes toward the end of training, as this is where the easiest and hardest instances are used for training. However, no such trend can be observed. Therefore, the first question is answered: there is no consistent impact of the second curricular block on the final result during the second half of training.

However, there is a trend in the local behavior at the beginning of the second half of the training. Similar to the observations in the first half, *h_r* induces the largest drop in averaged optimality gaps compared to the other curricular blocks in three out of four cases. In some cases, this goes so far that while *h_r* induces a sudden drop in the average optimality gap, the introduction of other curricular blocks invoke rises in the optimality gap.

A more fine-grained quantitative analysis of this observation, analyzing not only the averages over three seeded runs of the 16 curricula but all individual 48 runs, is shown in Fig. 5.8. The left side, Fig. 5.8a), shows how often each curricular block produced the largest drop in optimality gap in the first evaluation at 22,000 instances, right after its introduction, relative to the other curricular blocks. Rank one shows that *h_r* provoked the largest immediate drop in nine out of twelve cases. Analogously, rank four shows that *e_n* achieved the lowest performance in five out of twelve cases. The curricular blocks *e_n* and *h_n* do not show such strong trends. As before, it seems that training on particularly hard and easy instances generally induces a better and worse learning signal for the agent right at the beginning. In addition to the relative effects of curricular

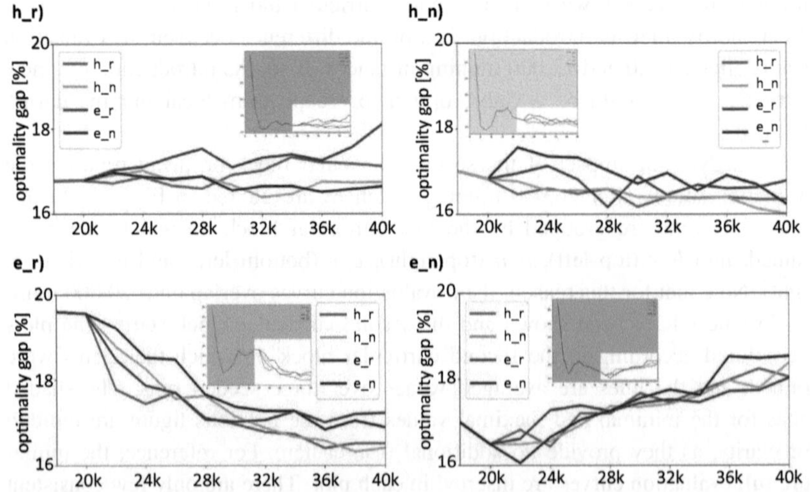

Fig. 5.7 Zoom-Ins on the second halves of agent training evaluation scores ordered by the common respective first curricular blocks

blocks, the absolute effects in terms of causing a local drop (positive impact) or rise (negative impact) in the optimality gap are depicted in Fig. 5.8b) and Fig. 5.8c). One can see a correlation that the harder the initial instances are, the more often a positive local impact is realized, and vice versa, the easier the initial instances are, the more often a negative local impact is observed.

This answers the second question, whether there is a consistent local impact when the second curricular block is introduced: The results are not consistent in that training on the hardest instances first (*h_r*) can lead to a worst local training effect at 22,000 instances and rises in the achieved optimality gap. Conversely, training on the easiest instances (*e_n*) in the middle of training can lead to the best local effect and local drops of the optimality gap. However, there is a very clear trend indicating that in general, training on a set of particularly hard instances in the middle of training induces a positive local learning signal and training on particularly easy instances is less favorable.

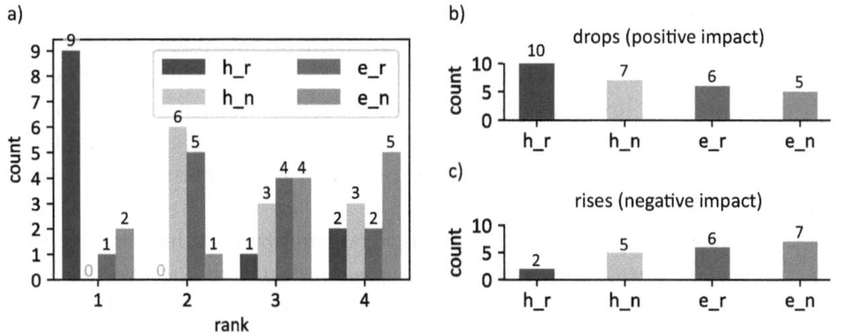

Fig. 5.8 Bar plots showing the relative and absolute immediate impact of introducing the second curricular blocks in training

5.4 Summary and Discussion

The study on CL for JSS in this chapter involved the creation of a novel measure of the difficulty of individual JSSP instances and different sequences in which instances can be ordered according to this measure to provoke different training effects. In contrast to existing difficulty measures which are mostly based on problem sizes, the presented approach takes the perspective of how well instances can be solved with known priority rules. It therefore proposes to use the makespan of the solution constructed from the MTR priority rule as an absolute measure of instance difficulty, such that instances with a large makespan are defined as hard and those with a small makespan as easy. To analyze the efficacy of this measure, a training dataset of JSSP instances was solved with MTR and sorted according to the achieved makespan. The sorted training dataset was then divided into easy and hard instances, and both halves were sorted either from easy to hard or from hard to easy, resulting in four curricular blocks of training instances. The pairwise combination of these curricular blocks resulted in sixteen curricula, on which the well-known DRL algorithm of Zhang et al. [67] was trained to observe the training behavior of the agents.

The results show that most agents reach their global minimum at the beginning of training. Here, training on the hardest instances first leads the agent to achieve the smallest makespan. Interestingly, the results are very consistent in that on average, the easier the instances are at the beginning of the curriculum, the better the agents perform. Moreover, training on any curriculum leads to better average makespans than training on randomly sorted instances. In the second half of the training, the agents converge towards a final performance that is worse than the one achieved early in training. Nevertheless, when transitioning from the first curricular block to the second, similar to the beginning of training, one observes that training on hard instances elicits an immediate positive response more often than training on easy instances. However, the effects of different curricular blocks in the second half of training are less consistent and less impactful with respect to the overall best performance.

The presented study makes several non-trivial contributions that shed light on RQ2, *how training strategies should be designed for efficient and effective DRL agent training.* Regarding the effectiveness of training strategies, which in the case of the JSSP means that the strategy steers the resulting agent policy towards achieving small makespans, the results in this study indicate that CL plays an important role. This finding is consistent with previous results on combinatorial optimization problems, as was presented in the related work in SubSection 5.1. However, a novel contribution is that the results of the introduced DTS measure indicate that there are measures beyond the problem size that can have a consistent effect on the learning behavior of DRL agents on scheduling problems. More specifically, it was shown that training on any curriculum that orders the training instances according to the DTS measure has a positive effect and is therefore preferable to a random training order. Moreover, starting training with harder instances leads to the best results in this case.

Regarding the efficiency of training strategies, the results in the presented study are more ambiguous. An efficient training strategy would yield equal or better results after less training time. However, the evaluation curves in the presented study show an atypical behavior where the agents reach a local minimum after the same number of training instances. Therefore, at the granularity level of 2,000 training instances, all curricula are equally efficient. However, it is conceivable to leverage the gained insight that training on particularly hard problem instances induces a strong positive learning signal in situations in which learning has stagnated due to the convergence to a local optimum. If one takes the whole computation time required for training into consideration, one could argue that solving and sorting the training instances imposes additional computation time compared to training on randomly sorted instances. Solving and sorting the 40,000 training instances took only about 8 minutes on non-parallelized hardware, which is a 15 % increase in computation time compared to the average training time of roughly 54 minutes. On an absolute scale, however, the additional time is neglectable for the given problem size.

Despite the contributions, there are several limitations to the study. One limitation results from only conducting experiments on the base model of Zhang et al.

[67]. This approach was the best candidate for CL studies at the time the experiments were conducted, as it was the best performing algorithm and showed very good generalization ability. However, as mentioned above, the atypical learning behavior with the consistent local, and in most cases global, minimum and subsequent increase before a gradual decrease in hindsight makes it less suitable for studies regarding the efficiency in terms of learning speed. The achieved performance improvement of 1.1 %p over the best-performing model was significant at the time, but is rather small compared to the relative improvements achieved by new state-of-the-art DRL models, such as the 8.9 %p by Iklassov et al. [74]. Experiments on the transferability of the method to more recent learning-based construction heuristics should therefore be conducted in future work.

Like the base model by Zhang et al. [67], recent approaches show good generalization ability. However, CL strategies could still benefit from an explicit application of transfer learning methods. Iklassov et al. [74] incorporate a rehearsal strategy by revisiting old tasks, in their case small problem instances, during training. More advanced regularization-based transfer learning methods [228] could provide an additional benefit and could also mitigate the observed effect in the learning behavior of the approach of Zhang et al. [67] that the model performance becomes significantly worse after initial improvements.

Another limitation is the choice to experiment only with 6×6 problem instances. This choice was made deliberately, since computation times increase non-linearly with larger problem size due to the quadratic growth of the graph representations and the corresponding NNs. As mentioned above, training a single agent took roughly 54 minutes on average, such that three seeded runs on 16 curricula and the additional runs without CL took a total of $54 minutes \times 3 \times 17 = 2{,}754 minutes \approx 46 hours$. Hence, running the same experiments on larger problem sizes quickly becomes intractable. Nevertheless, the effects of the proposed difficulty measure and the corresponding CL strategy could be less significant for larger problem sizes, where the distribution of processing times and job-machine constellations may become statistically more homogeneous due to more operations. In future work, it would therefore be worthwhile extending the experiments to larger problem sizes and to reduce the number of analyzed curricula to the extreme cases starting with e_n and h_r.

Finally, apart from the base model, transfer learning strategy and problem size, experimenting with only one novel difficulty measure limits the insights that can be gained regarding the question, why the introduced CL strategy evokes the observed effects. Evidently, solving instances with MTR and using the resulting makespan as difficulty measure to construct curricula works as CL strategy, but it is not yet clear why. However, given the simplicity of the MTR dispatching heuristic, it seems unlikely that the proposed measure is optimal. Nevertheless, if there exists a better or theoretically optimal difficulty measure, it seems that the proposed difficulty measure correlates with it. Testing further difficulty measures and their correlations to the proposed measure would offer more indications about what the optimal difficulty measure really is. Some rudimentary experiments have been conducted, in which problem instances were intentionally manipulated to increase the makespan that MTR (or any solution algorithm) would achieve. For example, the processing time of all operations in all problem instances of the training dataset was decreased or increased by 20 %, resulting in 20 % larger or smaller makespans, respectively and thereby larger or smaller DTS measures. Expecting that these more extreme cases would have an even greater CL effect, training runs with [*e_n*, *e_n*] and [*h_r*, *h_r*] were repeated. However, no such effect was observed.

Other ideas for CL remain to be explored in detail in future work. Perhaps the most obvious CL strategy would be to combine problem size-related curricula and the DTS measure, since curricula based on these two measures of difficulty have already been shown to have a significant positive effect on the training behavior. For example, one could follow the RASCL strategy of Iklassov et al. [74] for different problem sizes, but arrange same-sized training data according to the DTS measure. Each of the two CL strategies could also be taken to more extreme cases: generating more than 40,000 problem instances and applying the DTS measure would result in a greater number of particularly hard problem instances on which the DRL agent could be trained. Similarly, one could train agents on problem sizes other than just those in the Taillard benchmark [22], as done by Iklassov et al. [74], but train on even smaller and larger instances.

In conclusion, experiments by others and in this study indicate that CL plays an important role in designing efficient and effective learning strategies for DRL-based JSS, which serves as motivation for future work to explore this area even further.

6 Effective Inference Strategies with Trained Agents for the JSSP

The primary artifact of the development and training phases of DRL agents are trained NNs. For policy-based DRL algorithms specifically, the efforts in previous chapters focus on features space improvements and more efficient training curricula result in sets of parameters that define approximate policy functions. However, little research exists on advanced methods for the inference phase of trained NNs. Therefore, this chapter discusses two studies derived from two different perspectives that shed light on RQ3 of this thesis, namely, *how a trained DRL agent should be used, given a fixed computational budget during inference, to decrease the expected value of the minimal makespan of the generated solution*?

In the case of learned construction heuristics, the policy functions receive representations of intermediate scheduling states as input and produce action probabilities as output. The action probabilities can be interpreted as the learned probability that choosing the next operation of a particular job in the set of jobs will lead to the best final makespan if the current policy is followed.

During training, actions are typically sampled stochastically from the learned probability distribution. After training, during inference, NNs are usually deployed greedily. This means that the maximum value of the probability is chosen in each step. When solving a JSSP instance, this results in one deterministic solution with a fixed makespan.

While this inference strategy is straightforward and intuitive, it is not well aligned with the practical objective during the usage of the model. The practical objective is to strike a good balance in the trade-off between the solution quality, measured by a short makespan, and the computational time required to arrive at that solution. From this practical perspective, using models greedily leaves no

C. Waubert de Puiseau, *Advancing Training and Inference Methods for Reinforcement Learning-Based Job Shop Scheduling*,
https://doi.org/10.1007/978-3-658-51305-4_6

room for trading off more computation for shorter makespans. Recently, some researchers have reported results of sampling multiple solutions stochastically from the learned policy during inference [74, 182, 236]. Although their results indicate that the expected minimum makespan decreases as the number of solutions drawn from the learned policy increases, there has been no reported attempt to further improve the sampling strategy as a function of the computation time one is willing to spend on the solution generation. To fill the research gap, this attempt is made in Sect. 6.1 by introducing a method that aims to optimize the sampling process given a fixed number of solutions, and thereby indirectly a fixed computation time, to be sampled.

During the DRL training, the policy network converges towards a final policy. One can speak of successful training if a subjectively satisfactory average return is achieved. However, after the model has converged, it is by definition unlikely to surpass the performance threshold it converged towards. In other words, the influence of design choices for the DRL algorithm and the training procedure will have pushed the NN towards a local optimum and additional training time will not lead to the development of a better strategy if no substantial changes are introduced into the process. In Sect. 6.2, a method based on the Decision Transformer [237] is devised that aims at continuing training with a superior actor network that learns from and surpasses the quality of decisions of the originally trained DRL policy to arrive at an overall better performing model during inference.

6.1 Action Sampling in JSS

As outlined in the introduction to Chap. 6, the notion of action sampling refers to the process of repeatedly drawing samples from the learned policy of a trained DRL agent. Unlike deterministic approaches, where the argmax of the policy distribution is utilized at each step to construct a single solution, stochastic action sampling enables the generation of multiple, diverse solutions for a given JSSP instance. Action sampling is done under the assumption that learned policies are not perfect strategies that yield a global minimum for any problem, in which case generating one deterministic solution would suffice. The assumption is further, that a learned policy is rather able to generate solutions that lie in a favorable area of the solution space. This assumption leads to a useful visualization of the sampling process: sampling explores an area of the solution space close to the location of the deterministic solution in the post-training setting, enabled by a learned action prior parameterized by the NN. Importantly, sampling can thereby lead to further improvements with fixed parameters, after the parameter convergence during training through gradient descent is realized. Therefore, the probability of finding a superior solution increases with an increasing sample size.

In this subchapter, the ramifications of action sampling under a fixed computational budget are carefully analyzed, leading to the formulation of a novel approach for improving the action sampling process of a learned construction heuristic for the JSSP. The content of this subchapter is structured as follows: in 6.1.1, the theoretical foundation and intuition behind the action sampling method are given, reframing it as balancing between exploration and exploitation during inference. Section. 6.1.2 reviews relevant related literature on action sampling for the JSSP. The newly developed sampling method is detailed in Sect. 6.1.3 in conjunction with the experiments conducted to validate its efficacy. Afterwards, the experimental results are presented and discussed in Sect. 6.1.4. Finally, Sect. 6.1.5 summarizes and critically reflects the contributions and limitation of the results with respect to RQ3 of this thesis. Much of the content of this subchapter has been orally presented, discussed and published via the Workshop "Bridging the Gap Between AI Planning and Reinforcement Learning" of the International Conference on Automated Planning and Scheduling (ICAPS) 2024 [238].

6.1.1 Intuition

For the sake of clarity of the nomenclature in the subsequent sections, four types of sampling strategies from a categorical probability distribution that represents a multi-discrete action space will be distinguished hereon. They consist of the two relative directions on a spectrum between exploration and exploitation as well as two distinct points on that spectrum:

- *Deterministic Sampling* refers to the greedy application of argmax to the output policy logits and leads to a single deterministic solution for a given JSSP instance.
- *Stochastic Sampling* refers to sampling from a given multi-discrete probability distribution in which the likelihood of each action is given by the corresponding logit.
- *Exploitative Sampling* refers to sampling strategies, in which the likelihoods of actions are manipulated to increase the likelihood of already likely actions. This leads to sampled solutions that statistically stay closer to the deterministic solution in the solution space.

- *Explorative Sampling* refers to sampling strategies which assimilate the probability distribution to an equal distribution. This leads to greater variety in the sampled solutions, i.e., an increased probability of sampling solutions that diverge more from the deterministic solution.

As illustrated in Fig. 6.1, when taken to their limits, exploitative sampling converges towards deterministic sampling whereas explorative sampling converges towards sampling randomly from an equal distribution over all actions.

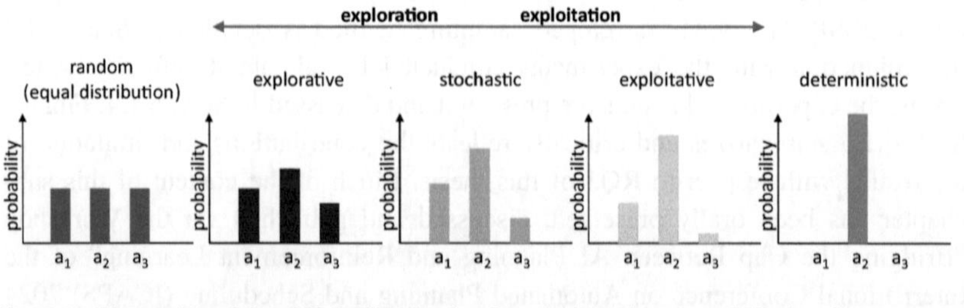

Fig. 6.1 Illustration of the shift in distribution from exploration to exploitation for an exemplary action space size of three

The fundamental intuition upon which action sampling methods build is that the expected minimal makespan among all sampled solutions of a problem instance m^* are likely to decrease (improve) with an increasing number of taken random samples β. Consequently, m^* is a function of β. Moreover, the steepness of this function changes as the probability distribution, from which actions are sampled, changes, as depicted in Fig. 6.2. While sampling deterministically from a trained policy is expected to yield a "good" solution, it will reproduce the same schedule every time and therefore will not improve through increased sampling. On the other hand, exploitative sampling will produce schedules close to the deterministic solution. These solutions may be worse than the deterministic solution for very small β-values but can lead to slightly better m^* for larger β-values through comparatively more exploration. Explorative sampling, on the other hand, broadens the range of different solutions that are found during sampling and enhances the relative effects: The generation of a single solution with increased randomness is likely to yield a more random and, consequently, less optimal m^*. However, for very large β-values, the sampled solutions will cover a wider area of the solution space and the chance of finding a better m^* increases.

If this intuition is valid and the functions of m^* over the sample size β look similar to the schematic plot in Fig. 6.2, this implies that, given a non-perfect policy, one should adjust the sampling procedure depending on the sample size one is going to implement during inference of the trained model. Hence, this subchapter examines the validity of this intuition and introduces a method that allows the user to define the degree to which the sampling process is conducted explorative or exploitative.

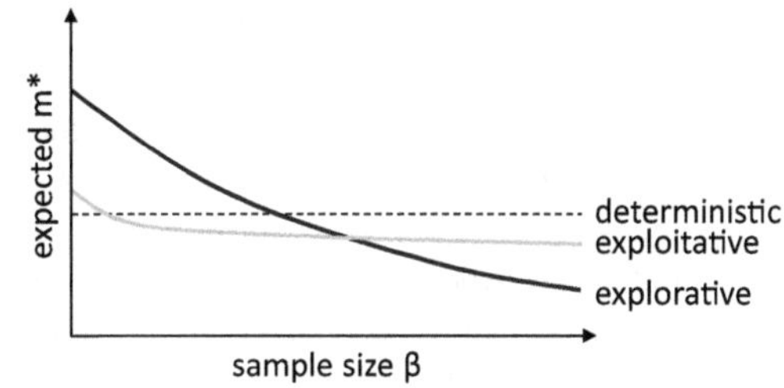

Fig. 6.2 Hypothesis on the effect of sampling exploratively and exploitatively with different sample sizes

6.1.2 Related Work: Action Sampling in JSS

Achieving an effective solution space coverage through sufficient exploration while experiencing enough exploitation to eventually converge towards a successful policy is an ongoing challenge when training DRL algorithms. Therefore, a plethora of sampling strategies has been developed, of which some are DRL algorithm specific, and others can be applied universally. One such strategy is epsilon greedy (ε-greedy) sampling, which involves drawing a random action from a uniform distribution instead of following the argmax of the learned Q-value with a probability of ε, thereby adding a random exploration factor to episode rollouts [44]. This method is common in value-based DRL algorithms and is effective. However, it disregards all learned priors in ε percent of the time, which can result in rollouts that diverge too much from learned strategy, hindering its efficacy during inference. In hybrid algorithms, such as Deep Deterministic Policy Gradient (DDPG) methods [54], Gaussian noise or noise drawn from an Ornstein-Uhlenbeck process [239] is added to the action vector to increase exploration. Policy-based methods, such as REINFORCE [49] and PPO [52], sample from the vector of action logits predicted by the actor network during training, which equates to the above definition of stochastic sampling. In certain implementations of PPO, the loss is regularized through an entropy term during training to push the agent behavior towards uncertain states [52], whereas other methods add related curiosity terms [240]. Although these methods work provably well for training, loss regularization terms cannot be used for forward rollouts during inference, which is the objective of this subchapter.

During inference, stochastic sampling is a common concept for guiding tree search algorithms in different domains [241]. The most well-known tree search algorithm in combination with DRL is the MCTS, which was famously leveraged in AlphaZero [1], as mentioned in the introduction of this thesis. In this method, the expansion of nodes representing states is determined by a linear combination of Q-values or action predictions and the visit count of nodes. The weighting of this linear combination allows users to steer the exploration-exploitation tradeoff, a powerful combination that has recently also been applied to related scheduling problems [182, 236]. Noteworthily, Shen et al. [242] and Wang et al. [243] have applied an exponent across the action vector within MCTS rollout generation to more smoothly steer the exploration-exploitation behavior. The primary effect that the authors utilize is the reduction or increase in the weight of the manipulated action vector relative to the visit counts of nodes. However, a secondary effect is the (dis-)alignment of values in the action vector itself. The method derived in this subchapter is inspired by this secondary effect. MCTS, however, incurs a high computational expense that is acceptable for the difficult task of evaluating states in stochastic environments, but otherwise inefficient. In contrast, for the deterministic JSSP, it is more efficient to sample multiple whole depth-wise trajectories from start to finish in parallelized code and the appropriate corresponding accelerated GPU hardware instead of deploying the inherently sequential MCTS.

Different forms of stochastic sampling during inference have, to a limited extent, also been examined and applied in learned construction heuristics for combinatorial optimization problems. Kool et al. [244] have shown the efficacy of stochastic sampling for DRL agents trained with REINFORCE on different routing problems. A derivative of REINFORCE called Policy Optimization with Multiple Optima (POMO) has been introduced by Kwon et al. [245], who in the same publication introduce a sampling routine which completes k greedy rollouts after k possible first action have been carried out to produce k distinct solutions. Although the authors demonstrate that this sampling routine works well on multiple combinatorial optimization problems, it is likely less effective for construction heuristics on the highly solution symmetrical JSSP. The solution symmetry, in this context, refers to the fact that different action sequences lead to the same schedule. A trivial example for this is if the operations of two different jobs need to be scheduled to different machines. The order in which these two operations are chosen to be integrated into an empty schedule is irrelevant, since the operations will be scheduled to time zero either way. Corsini et al. [56] train a construction heuristic for the JSSP with a self-supervised training routine and also confirm that stochastic sampling improves the obtainable makespans during training and test time. Pirnay et al. [57] concurrently developed a similar self-supervised training routine for the Travelling Salesman Problem and the JSSP, but additionally contribute an improved and more sophisticated sampling process. This process combines three techniques: beam search and top-p sampling (also known as nucleus sampling), which are both common techniques for sampling diverse sequences (sentences) from large language models [246]; and thirdly a newly developed technique, which they introduce as "sampling without replacement, but improvement". Beam search is a breadth-first algorithm which expands all leaf nodes in each step and then prunes those branches with the smallest predicted action probability of the subsequent step until a fixed number of parallel branches, referred to as beam width, remains [247]. Top-p sampling, in contrast, sorts the action probabilities in each step by size and expands the nodes from largest to smallest action probability until the sum of the probabilities of expanded nodes reaches a threshold value "p" [246]. "Sampling without replacement, but improvement" refers to a sampling process that produces samples in multiple sequential rounds to build a search tree, and after each round manipulating the original action probabilities predicted by the network depending on how successful the branch beyond an action is [57]. The authors demonstrate that the sampling process is superior to stochastic sampling during both training and inference. However, it requires sampling in multiple sequential rounds, which makes it slower. Although the results of Pirnay et al. [57] and Corsini et al. [56] are very promising for self-supervised models, it has not been studied whether the sampling strategies transfer well to DRL-trained models, which often suffer from policies with lower entropy.

Different sampling strategies have, however, also been applied to DRL-trained models. Tassel et al. [248] train a DRL-model on large-scale scheduling problems and during inference create multiple samples in the following way: after the creation of a greedy deterministic schedule, they re-visit those nodes of the corresponding tree that showed the largest uncertainty in the action prediction, measured by entropy, and also expand the node by taking the second most likely action, before continuing greedy expansions from there on. Unfortunately, the authors do not compare this strategy to other strategies. Since it is, for its sequential nature, not well parallelizable one can assume that it is slower than other methods. Therefore, it is not included in the study of this subsection. However, the assumption should be experimentally verified in future studies. Iklassov et al.

[74], on the other hand, analyze the performance of trained DRL agents with different sampling strategies, stochastic sampling, beam search and the sampling strategy introduced alongside POMO. In their study, stochastic sampling outperforms the other sampling strategies by a wide margin. Based on these results, beam search and POMO are not compared to the results in this subsection.

In conclusion, while various inference techniques have been introduced in the past, stochastic sampling stands out as a straightforward, highly parallelizable and effective method that works well for the JSSP in many cases. However, no method so far explicitly steers the exploration-exploitation behavior during inference to improve upon stochastic sampling.

6.1.3 Experiment Design

To steer the exploration-exploitation behavior during sampling, this study proposes the introduction of a temperature parameter δ that is applied to the action vector of a trained model a as in equation (6.3). This vector manipulation exponentiates every predicted output logit with the exponent δ and re-scales the vector, such that the sum of all values it contains equals one.

$$a_{sample} = \frac{a^{\delta}}{\sum_i a_i^{\delta}} \tag{6.1}$$

The effect of the manipulation on sampling is that for $\delta > 1$, relative differences between logits are emphasized, favoring exploitation, whereas for $\delta < 1$ those differences are decreased, leading to more explorative behavior. For $\delta = 1$, all logits remain unchanged, defaulting to stochastic sampling. Sampling from action vectors that are manipulated in this way is referred to as δ-sampling from here on. The δ-sampling method has the following advantageous properties:

- It allows a smooth and precise adjustment of the action vector and corresponding exploration and exploitation behavior of the trained agent.
- It easily scales to any action vector size and thereby to any JSSP instance size.
- It is a single-parameter method, such that fine-tuning it for a given trained policy and acceptable number of samples (β) does not become too computationally expensive.

Given a trained policy as base model and an acceptable number of samples β, a method is required that returns the δ* that minimizes the expected minimal makespan within all taken samples m*. Assuming that extreme exploration and exploitation are suboptimal, it is hypothesized that the function of m*(δ*) is smooth and possesses a single minimum at the point where the optimal balance between exploration and exploitation is achieved. This hypothesis motivates the search algorithm for δ* proposed here. It involves a grid search that iteratively and systematically decreases the distance between the two most promising candidate values for δ* to neighboring candidates by 50 % until a specified target distance between adjacent candidates is attained. Note that although one theoretically expects a single minimum and searching the periphery of a single most promising candidate should suffice, in practice, searching the periphery of two most promising candidates proved to provide more robust results. This is due to the fact that the search procedure involves the statistical evaluation of δ*-values on a finite evaluation data set, and the optimal value of this subset may statistically deviate from the globally optimal value. An illustrative example of a δ*search is presented in Fig. 6.3.

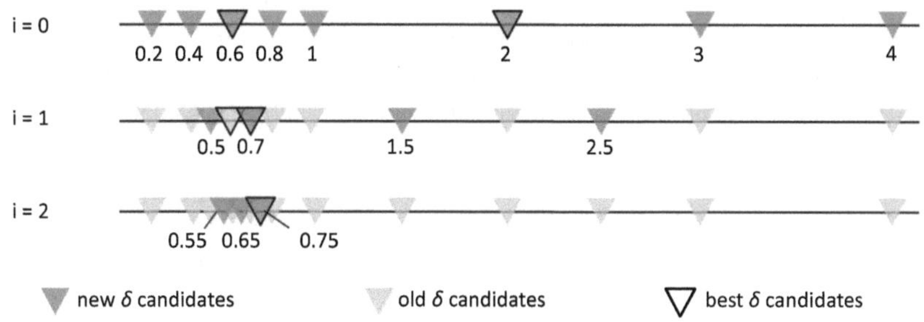

Fig. 6.3 Three iterations of the proposed search algorithm for δ^*

In the first iteration (i = 0), a predefined set of candidate δ-values is applied to sample β solutions on a set of evaluation instances. The exemplary initial candidate δ*-values {0.2, 0.4, 0.6, 0.8, 1, 2, 3, 4} are depicted as triangles along a horizontal linear scale. In the given example, δ = 0.6 and δ = 2 return the smallest makespans as indicated by the bold edges on these triangles. For the next iteration (i = 1), new candidates are obtained around the two best candidates {0.6, 2} from the previous iteration (i = 0). To this end, half of the distances to each adjacent candidate are subtracted from or added to each respective candidate, resulting in up to four new candidates for the second iteration (i = 1). These new candidates represented by new green triangles. Old candidates are gray. Again, these new candidates are used for δ-sampling. Within the list of all candidates,

$\delta = 0.6$ and $\delta = 0.7$ now yield the smallest makespans. Again, new candidates are added half the distance between the best candidates and adjacent counterparts for the third iteration ($i = 2$). Here, only three new candidates emerge, because the best candidates of the second two are adjacent to one another. Iterations are terminated once the largest distance between existing candidates and a new candidate falls below a predefined threshold. The algorithm then yields the δ-value with the smallest found makespan as δ^*.

In the following, the effect of δ -sampling is examined on two distinct DRL-based JSS algorithms that have been proposed. The first is taken from Zhang et al. (2020) [67], the most referenced algorithm in the literature review conducted in Chap. 3, and the currently best performing algorithm introduced by Iklassov et al. (2023) [74]. For brevity, these two algorithms are henceforth referred to as L2D (Zhang et al.) [67] and L2G (Iklassov et al.) [74] in reference to their original publication titles. For L2D, the parameters of separately trained JSSP-size invariant GNN for problem sizes 6 × 6, 15 × 15, 20 × 20, and 100 × 20 are downloaded from the openly accessible code repository [249]. It should be noted that the parameters used here for the problem size 100 × 20 are obtained through training on 30 × 20 JSSPs, as done in the original publication [67]. For the experiments on the L2G algorithm models that use a set2set recurrent layer [250] to be JSSP-size agnostic, no trained network parameters are publicly available. Consequently, three models are trained according to the reported protocol, employing the following CL settings proposed in the original publication [67], which led to the best reported results for the respective test problem sizes: for the 15 × 15 and 6 × 6 JSSP models, adaptive curriculum learning [74] is applied across 6 × 6 and 15 × 15 JSSP instances. For the 20 × 20 and 100 × 20 base model, the RASCL method [74] is applied and the models are trained on 6 × 6, 15 × 15, and 20 × 20 JSSP instances. The remaining training settings are consistent with those documented by Iklassov et al. [74].

The two base models are created for and tested on the problem sizes 6 × 6, 15 × 15, 20 × 20, and 100 × 20 for the following reasons: 15 × 15 and 20 × 20 are common problem sizes in popular public benchmark datasets. 6 × 6 and 100 × 20 present the smallest and largest problem sizes in the most widely used public benchmarks, thus serving as adequate representatives of both magnitudes. Other common problem sizes (cf. Table 2.1) are deliberately excluded to limit the necessary computational expense of the experiments. Separate datasets are generated for training and testing purposes.

6.1.4 Results

Intuition Validation

As a preliminary analysis, the intuition that the best trade-off between exploration and exploitation during the inference changes when the target sample size changes is examined. To this end, ten random 6×6 JSSP instances are generated and the L2G base model is employed to sample 10,000 solutions (m^*_{10k}) per problem instance with four δ-sampling parameterizations:

1. Random sampling, which corresponds to $\delta = 0$ but is implemented as drawing valid actions from a uniform distribution;
2. $\delta = 1$ as baseline of how stochastic sampling is usually performed in the original papers;
3. $\delta = 0.05$ as extreme example of explorative sampling;
4. $\delta = 10$ as extreme example of exploitative sampling.

The results are plotted in Fig. 6.4, illustrating the average minimal makespan m^* that is achieved over different sample sizes on a logarithmic x-axis. To make up for statistical outliers prevalent in smaller sample sizes s, $m^*(s)$ is averaged across multiple hypothetical sampling runs drawn from the 10,000 solutions. Using Python array slicing notation, this can be formalized as in Equation (6.2):

$$meanm^*(s) = \frac{s}{10,000} \sum_{i=0}^{10,000/s} min\big(m^*_{10k}[i : s(i+1)]\big) \tag{6.2}$$

In Fig. 6.4, the x-axis is segmented into four distinct zones at line intersections of interest. In zone A, which in this example exclusively contains $samplesize = 1$, strongly exploitative sampling yields the best solution. In zones B and C, stochastic sampling explores the solution space more effectively than strongly exploitative sampling. Starting in zone D, explorative sampling outperforms less explorative sampling techniques. However, within the plotted region, explorative sampling is still more effective than random sampling from an equal distribution. Overall, Fig. 6.4 closely resembles Fig. 6.2 and thereby substantiates the intuition that the optimization of sampling given a trained model and sample size is attainable.

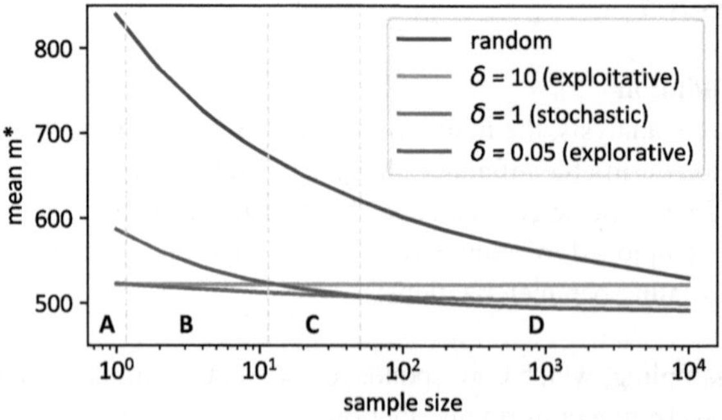

Fig. 6.4 Minimal makespans over sampling sizes for different sampling strategies

Optimal δ-Values

This next section presents the results of the search algorithm for finding and using the best δ-values for action sampling for sample sizes 32, 128 and 512. The selection of sample size 128 aligns with the sample size of the original L2G model [74]. Sample sizes 32 and 512 are selected as notably lower and larger exemplary sample sizes, respectively. It should be noted that on 100 × 20 JSSP problems with 512 samples are omitted, due to the expectation that they would contribute comparatively little additional insights compared to the considerable additional computational times required for solving such large instances. More background on this choice is presented in the discussion of this chapter (Subchapter 6.1.5). Table 6.1 summarizes the resulting best δ-values. Two general trends are observable. Firstly, in accordance with the aforementioned hypothesis, larger δ-values are consistently favorable for smaller sample sizes. Secondly, it appears that more exploitative sampling tends to be advantageous as the problem size increases for each base model.

However, there are also exceptions to these two general trends. For L2G and problem sizes 15 × 15 and 20 × 20, a modest increase in the found δ-values from sample size 32 to 128 is evident, followed by a subsequent decrease for sample size 512. Additionally, for L2G, the trend of increasing δ-values with increasing problem sizes discontinues for problem size 100 × 20. Zooming out, the found values for the L2G model are larger smaller compared to those found for the L2D model. Importantly, these discontinuing trends, exceptions, and differences between the base models suggest that analytical approaches to determining δ-values are intractable and they underscore the necessity to determine the δ-values empirically and individually for any trained model and sample size, as was done in this study.

Table 6.1 Best found δ-values for different base models, problem sizes and sample sizes

base	problem size	sample size		
		32	128	512
L2G	6×6	0.30	0.10	0.11
	15×15	0.70	0.86	0.43
	20×20	1.25	1.46	0.80
	100×20	0.58	0.43	n.a.
L2D	6×6	2.06	1.37	0.93
	15×15	12.25	6.02	3.32
	20×20	10.35	10.38	10.05
	100×20	14.00	10.88	n.a.

To gain deeper insights into the effect of δ-sampling, the mean m^* values obtained for various δ-values during the search of δ* are plotted for notable examples in Fig. 6.5. Across all examples, the function *meanm**(δ) appears to be relatively smooth and to have a single minimum, as was previously assumed. Nevertheless, some statistical noise prevails. Further, it seems like the iterative grid search method for δ* successfully converges towards reasonable global minima that are marked by the green semi-transparent triangles.

Upon closer inspection, Fig. 6.5 a) reveals the presence of multiple local minima for δ-values between 0.9 and 1.8 with m^*-values that are close to the globally minimal m^*-value. These findings imply that gradient-based methods for finding optima would be ineffective. Figure 6.5 b) further exemplifies a substantial plateau of m^*-values across a wide span of δ-values. In light of these insights, it is plausible that the sporadically observed outliers to trends in Table 6.1 discussed above are attributable to statistical noise.

Numerical Results of δ-sampling

The effectiveness of the obtained δ-sampling parameterizations is tested on 100 separate test instances. The average m^*-values for δ-sampling, stochastic sampling as baseline, and the deterministic solutions are presented in Table 6.2, along with the relative improvement between δ- and stochastic sampling calculated as in equation (6.3).

$$improvement = \frac{deterministic - \delta\text{- sampling}}{\delta\text{- sampling}} \tag{6.3}$$

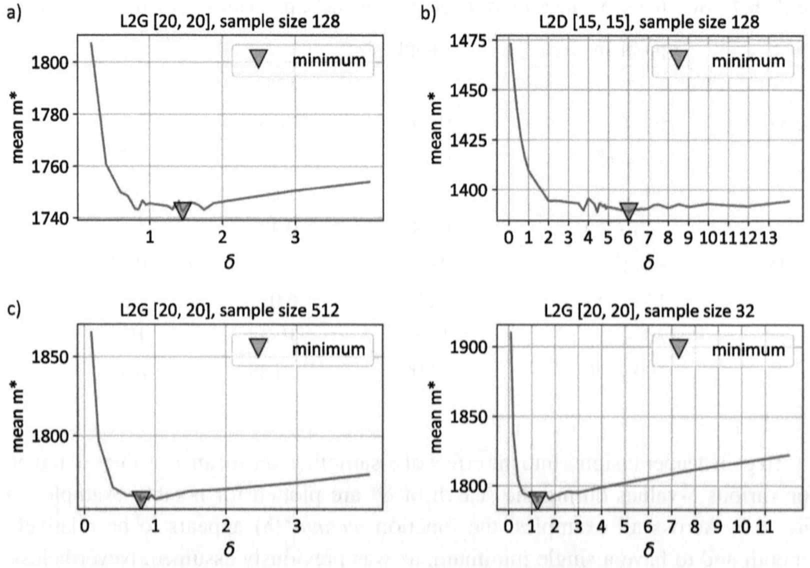

Fig. 6.5 Examples of results of the δ-value search algorithm with found minima

In the column entitled "improvement", positive improvements through δ-sampling are printed in green whereas negative improvements are printed in red. As anticipated, both sampling methods demonstrate superior performance in comparison to the deterministic solution, confirming previous results [74, 241, 244]. Moreover, -sampling outperforms the stochastic baseline in 18 out of 22 cases, achieving up to 3.4 % relative improvements. Notably, such improvements are realized across both models and across all sample sizes. The most notable exception lies in the worse result through δ-sampling with a sample size of 32 in L2D with 6 × 6 problem instances, where δ-sampling performs 0.6 % worse than stochastic sampling. This deviation can be partially attributed to the small sample size, which favors statistical outliers. The exception that persists across different sample sizes is in the results on L2G and 20 × 20 problem instances, in which δ-sampling consistently performs slightly worse than the stochastic baseline. For all three cases, the impact of changing δ-values is depicted in Fig. 6.5 a), c) and d). It is clearly visible that the minima of the function mean $m^*(\delta)$ reside in close proximity to one and within small plateaus that also include one, i.e. stochastic sampling. This implies that the δ search method is effective in finding stronger required manipulations where they are useful. Conversely, in scenarios where δ-values are close to one and are situated withing plateaus of the function mean $m^*(\delta)$, only minor and potentially worse sampling behavior differences are possible.

A question that arises when δ-values close to one and very similar m^* are obtained is whether the resulting schedules are the same. A particularly striking example is L2G on the 15 × 15 problem instances with a sample size of 128. In this case, a relative difference of less than 0.00 % between the average m^* obtained through δ-sampling ($\delta = 0.86$) and stochastic sampling is observed. Surprisingly, a pairwise comparison between m^*'s found across the 100 test instances showed that only 4 out of 100 instance solutions achieved the same m^*. This shows that even the δ-value close to one had a significant impact on the scheduling behavior. In contrast, on 6×6 JSSP instances solved with L2G and a sample size of 128, for roughly one third of the test instances the different sampling strategies resulted in the same m^*. This is most likely an effect of the smaller total solution space that is explored during sampling for instances of the smaller problem size. Despite the limited number of different solutions found, these solutions exhibit a substantial impact resulting in a 1.9 % improvement. This also underscores that δ-sampling is effective for problems with smaller solution spaces.

For the sake of better comparability, the obtained δ-sampling strategy is also applied to the TA benchmark instances [22] of the respective analyzed problem sizes. The results are printed in Table 6.3. In comparison to the results on the self-created random JSSP instances, there are only 10 instances per size. However, because optimal solutions are known, the optimal m is given in the column *opt.* and percentual gaps to the optimal m, are presented in the respective *gap* columns. The observable effects of δ-sampling are fundamentally similar to those described above for the self-created problem instances. While the results on L2D indicate even greater improvements through δ-sampling, the results on L2G are more ambiguous, both of which can be attributed to a larger statistical error. In absolute terms, it should be noted that for L2G, slightly larger makespans are found than those reported in the publication [74], as the reported performance was not reproduced by the self-trained models.

However, the most interesting insights this study is aiming to generate are the relative improvements between sampling methods. To this end, it is beneficial to visualize the tabular data in Table 6.3 as achieved m^* over different sample sizes, as done in Fig. 6.6. This facilitates the comparison of improvements achieved through increasing sample sizes and those achieved through better sampling (δ-sampling). Note that sample sizes are roughly proportional to computational time. In many cases, it is evident that the achieved m^* through δ-sampling is similar or better than the m^* achieved via stochastic sampling with four times larger sample sizes. In

Table 6.2 Average m^*-values on 100 generated test instances per problem size

sample size	base	problem size	δ–sampling	stochastic	deterministic	improvement
32	L2G	6 × 6	496,0	505,5	517,0	**1,9 %**
		15 × 15	1310,4	1311,4	1364,5	**0,1 %**
		20 × 20	1782,9	1782,3	1866,2	**0,0 %**
		100 × 20	5759,6	5764,1	5867,3	**0,1 %**
	L2D	6 × 6	519,3	516,1	577,1	**−0,6 %**
		15 × 15	1406,9	1434,6	1534,7	**1,9 %**
		20 × 20	1891,4	1957,4	2012,5	**3,4 %**
		100 × 20	5865,8	5928,4	5996,7	**1,1 %**
128	L2G	6 × 6	491,7	503,6	517,0	**2,4 %**
		15 × 15	1300,4	1300,9	1364,5	**0,0 %**
		20 × 20	1769,3	1767,6	1866,2	**−0,1 %**
		100 × 20	5733,6	5745,1	5867,3	**0,2 %**
	L2D	6 × 6	504,8	506,2	577,1	**0,3 %**
		15 × 15	1385,7	1410,1	1534,7	**1,7 %**
		20 × 20	1865,8	1927,6	2012,5	**3,2 %**
		100 × 20	5869,8	5899,7	5996,7	**0,5 %**
512	L2G	6 × 6	489,9	501,6	517,0	**2,3 %**
		15 × 15	1281,6	1291,7	1364,5	**0,8 %**
		20 × 20	1752,0	1750,6	1866,2	**−0,1 %**
	L2D	6 × 6	498,0	498,7	577,1	0,1 %
		15 × 15	1369,1	1386,9	1534,7	1,3 %
		20 × 20	1842,2	1879,6	2012,5	2,0 %

practical terms, this means that the same result can be achieved with a quarter of the necessary computational budget (or computational time), rendering δ-sampling four times more efficient (or faster) during inference.

Table 6.3 Average m^*-values on TA benchmark instances

s.–size	base	p.–size	δ–sampling		stochastic		deterministic		improv.	opt.
			m*	gap	m*	gap	m*	gap	in %	m
32	L2G	15 × 15	1345,9	9,5 %	1345,1	9,5 %	1417,4	15,3 %	**−0,1 %**	1228,9
		20 × 20	1858,9	14,9 %	1858,7	14,9 %	1935,3	19,7 %	**0,0 %**	1617,3
		100 × 20	5759,7	7,3 %	5784,6	7,8 %	5878,4	9,6 %	**0,4 %**	5365,7
	L2D	15 × 15	1452,5	18,2 %	1465,9	19,3 %	1530,5	24,5 %	**0,9 %**	1228,9
		20 × 20	1953,2	20,8 %	2027,1	25,3 %	2081,8	28,7 %	**3,6 %**	1617,3
		100 × 20	5928,6	10,5 %	5983,1	11,5 %	6089	13,5 %	**0,9 %**	5365,7
128	L2G	15 × 15	1340,3	9,1 %	1340,9	9,1 %	1417,4	15,3 %	**0,0 %**	1228,9
		20 × 20	1846,4	14,2 %	1837,4	13,6 %	1935,3	19,7 %	**−0,5 %**	1617,3
		100 × 20	5751,1	7,2 %	5766,5	7,5 %	5878,4	9,6 %	**0,3 %**	5365,7
	L2D	15 × 15	1416,2	15,2 %	1442	17,3 %	1530,5	24,5 %	**1,8 %**	1228,9
		20 × 20	1943,4	20,2 %	2002,7	23,8 %	2081,8	28,7 %	**3,0 %**	1617,3
		100 × 20	5888,2	9,7 %	5930,4	10,5 %	6089	13,5 %	**0,7 %**	5365,7
512	L2G	15 × 15	1323,1	7,7 %	1333,7	8,5 %	1417,4	15,3 %	**0,8 %**	1228,9
		20 × 20	1825,5	12,9 %	1828,9	13,1 %	1935,3	19,7 %	**0,2 %**	1617,3
	L2D	15 × 15	1409	14,7 %	1427,6	16,2 %	1530,5	24,5 %	**1,3 %**	1228,9
		20 × 20	1895,3	17,2 %	1946,2	20,3 %	2081,8	28,7 %	**2,6 %**	1617,3

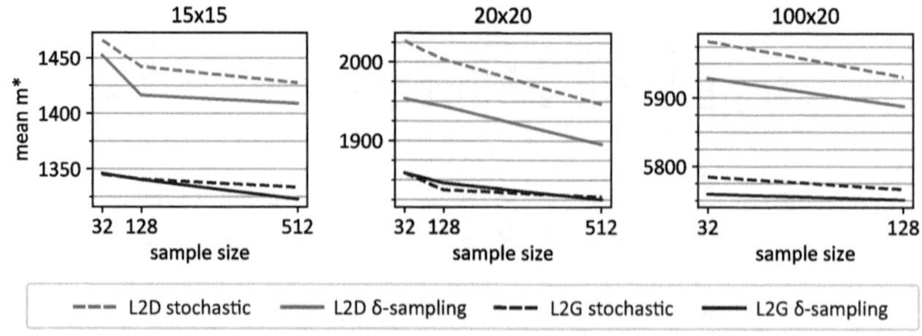

Fig. 6.6 Comparison of sampling methods over sample sizes

6.1.5 Summary and Discussion

The study conducted in this subchapter builds upon the intuition that a wider solution space can be explored when the available computational budget and corresponding feasible sample size is larger, and that this may lead to better solutions obtainable from a trained DRL agent during inference. The introduced δ-sampling seems to be effective in steering the sampling process of learned construction heuristics for the JSSP and can indeed be used to improve the expected makespan considerably when the right δ-values are used. This finding validates the underlying intuition. The introduced method employed for finding suitable δ-values also appears effective.

Through these findings, the presented study makes several valuable contributions to RQ3, *how a trained agent should be used, given a fixed computational budget, to decrease the expected value of the minimal makespan found during inference*. The first and most general contribution lies in the insight that the targeted, sample-size-dependent adjustment to the exploration-exploitation tradeoff during inference has a positive impact on the expected value of the minimal makespan. In general, promoting exploration when larger problem sizes are acceptable leads to better results. The second contribution is the introduction of δ-sampling, a method that allows users of a trained model to adjust the exploration-exploitation behavior smoothly and effectively in a straightforward manner through a single parameter. The third and most specific contribution is the development of a suitable algorithm to find the right δ-value effectively and in accordance with the properties of the function underlying the performance in dependence of the δ-value, i.e., being smooth and having a single minimum.

Although the study was only conducted on two base models for the JSSP, it is reasonable to assume that the described intuition and methods can transfer to other learned construction heuristics. Moreover, these methods could also be applicable to other related combinatorial optimization problems, such as the Travelling Salesman Problem or the Capacitated Vehicle Routing Problem, for which constructive approaches exist.

However, there are also limitations and potential improvements regarding both the study design and the suggested methods. One potential limitation is that both chosen base models are trained through DRL to find one good solution through stochastic sampling. Although this is the most common way, it is conceivable that agents could in theory be trained to create sets of multiple solutions in a MCTS-like forward simulation and be rewarded with respect to minimizing the makespan of one solution within each set instead of maximizing the expected value of a single strategy. Although none of the studies analyzed in Chap. 3 make use of such a method, one should note that δ-sampling may be less effective in such scenarios. This is because δ-sampling operates under the assumption that the computational budget during inference is unknown during training and the agent has not been fine-tuned towards it.

Regarding the study design, another limitation is that experiments involving a sample size of 512 on 100×20 problems were omitted. As previously mentioned, it is improbable that qualitatively different insights would have been generated, though this cannot be entirely ruled out. However, the required computational times for finding suitable δ-values is substantial. To illustrate his point, it should be considered that testing 21 δ-values with 128 samples on 100 15×15 JSSP instances for the much faster L2G model took around seven hours on the used GPU-accelerated hardware. If the sample size was fourfold and the number of actions required per instance roughly ninefold, the testing time would exceed ten days.

Another limitation lies in the δ-sampling method itself, which, as was outlined in the related work of this subchapter, is not the only way to steer the exploration-exploitation behavior of a trained model. Especially MCTS and "sampling without replacement, but improvement" in multiple rounds [57] are promising candidates for a targeted exploration of the solution space. The δ-sampling method is most efficient, when all solutions can be created in parallel on suitable hardware. Hence, the aforementioned alternatives have the biggest advantage, if batches have to be processed sequentially on the available hardware. It is a promising direction of future work to explore these more sophisticated inference techniques and the development of techniques to efficiently find suitable parameterizations of them.

Finally, it was observed that the devised algorithm for finding δ^* is effective. However, it was also observed that large plateaus of resulting *meanm** can form, in which the proposed algorithm can lead to many unnecessary and expensive sampling trials. With the newly gained insights, the algorithm can likely be improved in terms of efficiency in the future. For example, one could exclude δ-values that, upon the identification of a plateau, are evident to clearly lie outside of the plateau. Alternatively, the deployment of early stopping when only minimal changes in the resulting *meanm** can be expected from a continued search could be considered in future work.

6.2 Decision Transformers for the JSSP

The previous Subchapter 6.1 explores a method to increase the likelihood of sampling a particularly good solution among multiple solutions of a JSSP instance after a model has been trained to convergence. In other words, computational budget is spent on finding a better solution *during inference*. This subchapter takes a different approach. It introduces a method to train on, learn from and improve upon the policy of a converged DRL agent. This allows spending further computational budget on training as a *subsequent* step after the initial DRL training has converged, whereas the computational budget for inference remains untouched. Specifically, this method makes use of the recently introduced Decision Transformer (DT) paradigm [237], which exploits the strong expressiveness of the Transformer architecture [251] on sequential data to solve sequential decision-making problems in a DRL-like fashion.

Essentially, the DT model takes as input at time t the current state S_t, the expected sum of all future rewards, called return-to-go R_t, and the recent history of return-to-go–state–action tuples $(\ldots R_{t-1}, S_{t-1}, A_{t-1}, R_t, S_t)$ to predict the next action A_t. The training is conducted in a supervised manner on a generated dataset of state-action-reward tuples. A more detailed introduction to the DT is provided in subchapter 6.2.2. In contrast to the DT, a typical DRL model takes the current state s_t as input to predict an action a_t, here noted in lower-case for easier differentiation to the states and actions of the DT. Because of the way in which the DT is trained, it has two potential advantages over the trained DRL model. First, by considering a history of past actions at each step, the DT can learn how previous actions in the same episode, i.e., on the same problem instance, have influenced the solution and can adjust its strategy accordingly. Second, the DT is trained to adjust its actions in such a way as to achieve the return-to-go it is given as input at any point in the decision-making process. Therefore, a particularly challenging return-to-go prior may push it to improve its solution strategy and lead to performance improvements over the strategy used to generate the training dataset. Although the DT could be trained on data collected from a random policy, it has been shown to be more efficient when trained on data of successful trajectories [237]. In such cases, one can speak of the DT as "student" and the strategy used for training data generation as "teacher".

Motivated by these potential benefits, this subchapter investigates the hypothesis that by training a DT on a dataset generated through the interaction of a trained DRL model for JSSP with its environment, one can create a model that outperforms the trained DRL model during inference. Thereby, this subchapter sheds light on another new method related to RQ3, *how a trained DRL agent should be used, given a fixed computational budget during inference, to decrease the expected value of the minimal makespan of the generated solution*?

To this end, this subchapter is structured as follows: in 6.2.1, the related work of DTs for the JSSP discusses successful studies on the application of DTS in other domains, student-teacher learning methods and their application to the JSSP, and a set of methods called "active search" for combinatorial optimization problems. Then, in Sect. 6.2.2, the experiment design is described. This includes an introduction to the DT method, the DRL model used as baseline and for the generation of the expert trajectories in a training dataset, which is referred to as *teacher model*, and the methods and study design. Section 6.2.3 presents the experimental results, before they are discussed with respect to contributions to RQ3 and their limitations in Sect. 6.2.4. The results of this chapter have been published in [252].

6.2.1 Related Work: Decision Transformers for the JSSP

DTs belong to the class of offline RL algorithms that devises machine learning models to learn from static datasets consisting of sequential decision-making data [253], i.e., state-action-reward tuples. Recent well-known methods include Conservative Q-Learning [254], Implicit Q-Learning [255] and Q-value regularized transformers [256]. These algorithms, as their names suggest, primarily learn to approximate Q-values from offline data and are reported to struggle with value overestimation and are sensitive to out-of-distribution data [257]. Furthermore, the literature review in Chap. 3 indicates that, at least in DRL-based methods, value-based approaches are less effective for scheduling problems than policy-based approaches. Although it is unclear whether this limitation applies to the offline RL algorithms in the specific setup presented here, these findings still motivate the preference of other offline RL algorithms over value-based ones in this study. For example, Fujimoto and Gu. [257] add a behavior cloning term to a policy-based RL algorithm. In Trajectory Stitching, introduced by Hepburn et al. [258], this idea is enriched by a model-based approach to augment data between known states. Compared to these two methods, the DT is simpler in its design, directly builds on the advances in large language model research by using the transformer architecture, and does not require a learned model of the environment [237].

In the context of the DT, one should also mention the Trajectory Transformer (TT) [259], which was developed concurrently with the DT and also uses the transformer architecture for action sequence modeling. The main difference between those two lies in the return-to-go prior of the DT, which is replaced by a goal-condition mechanism, and the inclusion of beam search to provoke a deviation from the strategy deployed in the dataset creation process. There seem to be no clear theoretical advantages or disadvantages of the TT over the DT. For the study in this subchapter, the DT was favored, since at the time of conducting the study, it had roughly twice the citation count, indicating a larger popularity. However, the TT remains an interesting alternative that could be explored in future work.

Since its introduction in 2021, the DT has been applied successfully to various problem settings, including in the medical context for scheduling doctor-patient visits and treatment plans [260], for placing circuit modules onto semiconductor chips [261], for active object detection in robotic control [262], and for favorable actions in e-commerce recommender systems [263]. The widespread success has motivated the development of advanced modifications of the DT. The Hierarchical DT [264] divides the learning problem into training a high-level agent that selects subgoals for a lower-level agent. The Graph DT [265] models the input sequence as a decision graph to facilitate learning the temporal and causal relationships in the data, which are otherwise learned through the self-attention mechanisms alone. The latest addition to the family of DTs is the Elastic DT [266], which improves the combination of suboptimal trajectories into a more favorable one through flexible context lengths. Although the extensions of the DT have shown superior performance on certain tasks, they introduce additional design choices and hyperparameters that need to be tuned. For the initial proof of concept of the methodology explored in this subchapter, the DT is therefore preferred over its successors to keep the computational cost of hyperparameter tuning small. However, they present a promising direction for future research. None of the above offline DRL methods have been applied to JSS prior to this study.

Through training a DT on a dataset that is generated by a trained DRL model, the method introduced in this subchapter is related to several areas of research. For example, Imitation Learning explores the concept of student-teacher learning, where a "student" model learns to replicate strategies from policy demonstrations of a "teacher" [267]. A common subcategory in this area is Knowledge Distillation [268], where smaller NNs are trained to mimic the learned outputs of larger NNs, typically to speed up inference at a tradeoff in accuracy [269]. Although this is an interesting related approach for JSS and has not been previously studied, the method introduced in this chapter aims to achieve the opposite: improve the strategy and accept a slower network inference. This direction is arguably more interesting, because as an engineer, one may be able to find useful patterns in a superior learned strategy that may be valuable for future designs, such as features or action spaces.

There is a limited amount of research on student-teacher-like learning routines to solve scheduling problems. An early example comes from Ingimundardottir and Runarsson [270] in 2018, who trained a linear machine learning model on data generated with dispatching rules and were able to replicate the set of rules. Rinciog et al. [162], for a JSSP-related scheduling problem, pre-trained an NN supervised on the earliest-due-date dispatching rule before fine-tuning it with DRL. Although these approaches showed some success of imitation learning, the results are far from competitive compared to other algorithms in terms of the achieved schedules due to their weak teacher algorithms. To address this shortcoming, Lee and Kim [271] trained a GNN on a set of optimal solutions. However, the learned policy also did not perform competitively compared to modern DRL-based methods, and the dataset generation is computationally expensive. Some authors argue that trying to learn from perfect solutions is infeasible for combinatorial optimization problems [56, 57]. In contrast, the method presented in this subchapter proposes to train on a dataset generated by an already competitive strategy DRL agent compared to other algorithms. This ensures that the teacher strategy is parameterizable through a DL model and makes it more likely that the resulting final policy is competitive as well.

Offline RL is not the only way to spend additional training time to improve a policy. In so-called Active Search (AS), untrained agents or agents that have converged during DRL training are trained on the problem instance one is trying to solve to effectively derive a problem instance-specific strategy [272–274]. AS is effective for several combinatorial optimization problems but has not been specifically optimized for the JSSP. It differs significantly from the method proposed in this study in that additional training time is spent at inference-time for a specific problem instance, rather than spending it during an additional dedicated training time separately and prior to inference. With respect to the research question posed in this chapter, AS is a further very interesting candidate for future research endeavors.

The application of the transformer architecture is ubiquitous in DL and DRL research domains, including work on scheduling problems. Bonetta et al. [66] and Zhao et al. [79] use the self-attention mechanism to capture the structure of the scheduling instance by treating each required operation as an input token. Pirnay et al. [57] introduce a more efficient alternative, in which self-attention is alternately applied between operations of the same job and operations that need to be processed on the same machine. Yet, in applying the DT to the JSSP, the transformer architecture is leveraged differently in the method proposed in this subchapter. Here, individual tokens are not individual operations, but entire solutions to a problem instance, action representations, or return-to-go representations.

In summary, the method described and analyzed in the remainder of this subchapter presents several novel elements: it is the first study of the application of an offline RL method on the JSSP, it uses a trained DRL agent as teacher model, and it leverages the transformer architecture to effectively encode and compare entire intermediate scheduling states.

6.2.2 Experiment Design

The methodological experiment workflow can be divided into three distinct phases, which are conceptually illustrated in Fig. 6.7. The figure outlines how the dataset generation, subsequent DT-Training and finally the DT-Testing are conducted, and which components are required to accomplish each step.

Before going into the details of the three successive steps, brief introductions to the DT and the trained DRL model used in this study are given. The introductions are intended to provide the necessary context and to introduce the notation used throughout this subchapter. The description of the DT focuses on the differences between the DRL and the DT learning paradigm, as the concepts and notation can otherwise easily be confused. For more details on the DT, the reader is referred to the original publication [237]. Similarly, the provided description of the used DRL model is intended to briefly summarize the information needed to understand the method introduced by Falkner et al. in [42].

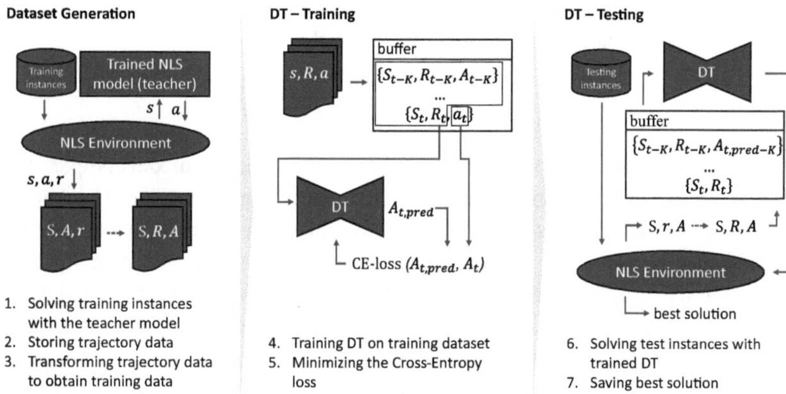

Fig. 6.7 Experiment workflow of the Decision Transformer study

Decision Transformers

The DT method was introduced by Chen et al. in 2021 as a way to leverage the successful transformer architecture to address MDPs as sequence modeling problems [237]. As such, like RL methods, it deals with states, actions, and rewards of an MDP, but in a fundamentally different way compared to DRL models. In RL, interaction sequences between the agent and the environment are lists of state-action-reward tuples, which can be formally expressed as:

$$sequence_{RL} = \left[\{S_0, A_0, r_0\}, \{S_1, A_1, r_1\}, \ldots, \left\{S_{final}, A_{final}, r_{final}\right\}\right], \tag{6.4}$$

where the environment provides the state information S_t at step t, the model returns the action A_t and receives the reward r_t. Therefore, simply put, the DRL deep learning model, $model_{DRL}$ approximates the function:

$$A_t = model_{DRL}(S_t) \tag{6.5}$$

In contrast, DTs take each action based on an input sequence of the last K states, actions, and return-to-go values R. In this context, K is referred to as the context length. The return-to-go at any given time step t, denoted by R_t (the uppercase R is used here to distinguish it from the lowercase r of the reward), is defined as the sum of the remaining rewards of the episode: $R_t = r_t + r_{t+1} + \ldots + r_{T-1} + r_T$, where T is the terminal state. Specifically, at each step t, the DT deep learning model,

$model_{DT}$, takes as input the last K states, actions and returns-to-go, along with the current state S_t and the current return-to-go prior R_t:

$$A_t = model_{DT}(R_{t-K}, S_{t-K}, A_{t-K}, \ldots, R_t, S_t) \tag{6.6}$$

In other words, the two major differences in the input of the DT model compared to the DRL model are the historical context and the return-to-go prior. The DT model is trained to generate action sequences that aim to achieve the target R_t by minimizing the difference between the target R_t and the cumulative sum of future rewards collected in the experience data. During inference, R is then set to a manual prior value R_{prior}, which can bias the DT to output action sequences that achieve a greater return-to-go than that observed in the training data $R_{observed}$, i.e., $R_{prior} > R_{observed}$. Note that the interaction data used as training data can come from different sources. It can, for example, be generated randomly, through expert actions or, as is done in this study, by a trained DRL agent. Since in the latter case, the DT "learns from" the policy of a DRL model, the DRL model will be referred to as *teacher model* from here on. As a sequential deep learning model, the transformer architecture [251] is used, which gives the DT its name. The data flow through the abstracted transformer is depicted in Fig. 6.8.

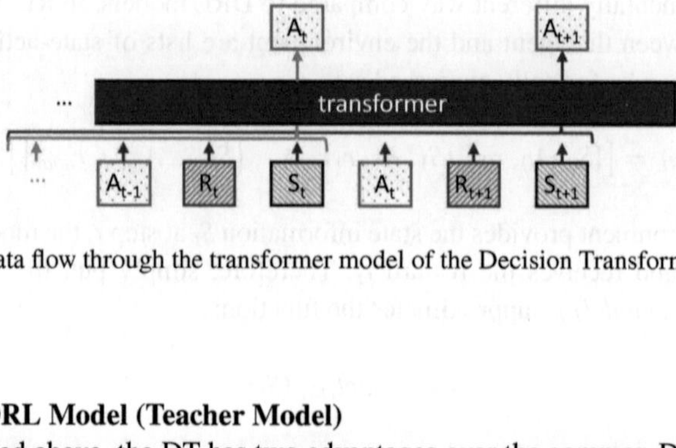

Fig. 6.8 Data flow through the transformer model of the Decision Transformer (inspired by [237])

Trained DRL Model (Teacher Model)

As described above, the DT has two advantages over the common DRL approach: The longer history of interactions that serves as context for the next decision, and the manipulable return-to-go prior. To effectively use the history of interactions, the interactions should contain useful information about the current problem instance. The learned construction heuristics that were discussed in previous chapters typically do not fulfill this property. This is because the only necessary information for a construction heuristic approach lies in the last operations processed by each machine, and information about the remaining unscheduled operations. Only this information dictates where the next unscheduled operations can be filled in. Any information on the history of actions, i.e., how the current situation was reached, is unnecessary.

For a teacher model in this application, one is looking for an approach in which past interactions with a JSSP instance could reveal instance-specific characteristics that may be leveraged. By this train of thought, a more promising class of DRL-based JSSP solution methods than learned construction heuristics are learned metaheuristics. A recent and very successful representative of this class is Neural Local Search (NLS), introduced by Falkner et al. [42]. As the name suggests, the approach controls a local neighborhood search heuristic on the JSSP with a DRL agent.

The underlying local search heuristic iteratively improves upon an initial solution. At each iteration, the heuristic

- accepts or declines the solution of the last iteration,
- then *either* chooses a new neighborhood operation that defines the next neighborhood, i.e., the next set of solutions to generate in the search,
- *or* chooses a perturbation operator to jump to a new area in the search space, where the search then continues.

Falkner et al. [42] translate the heuristic decisions into actions that may be made by a DRL agent, experimenting with three different action spaces that are defined by the following three sets of actions:

1. Acceptance decisions: the decision whether the last local search step is accepted or not:

$$A_A := \{0, 1\}$$

2. Acceptance-Neighborhood decisions: a tuple representing the above acceptance decision and the choice between four different neighborhood operations in the set of ϕ:

$$A_{AN} := \{0, 1\} \times \phi$$

3. Acceptance-Neighborhood-Perturbation decision: tuple representing the acceptance and neighborhood decisions plus a perturbation decision from the perturbation operator set ψ:

$$A_{ANP} := \{0, 1\} \times \{\phi \cup \psi\}$$

The set of neighborhood operations ϕ comprises: the Critical Time (CT) operator [275], which swaps adjacent nodes in a critical block of the graph representation of the JSSP; the Critical End Time (CET) operator [276], which swaps the nodes at the start or end of a critical block; the Extended Critical End Time (ECET) operator [277], which swaps nodes at both the beginning and the end of a critical block; and the Critical End Improvement (CEI) operator [278], which shifts a node within ta critical block to a new position within the same block.

In the NLS approach, the agent chooses between these actions based on state descriptions that are derived from a learned representation consisting of the problem instance and its current solution. Falkner et al. [42] employ an encoder-decoder architecture which is depicted in Fig. 6.9. At a high level, the encoder uses a GNN to process the raw graph representation of the problem instance (cf. Subsection 2.1.2, Formalization and Characteristics of the JSSP, for more details). The aggregation part of the NN combines this representation with descriptive state features of the current solution, before the decoder translates the aggregated information into Q-values. These Q-values are learned through Double Deep Q-Learning [48] and aim to maximize the total improvement in the makespan between the initial and final solution of the JSSP instance across the episode. To this end, the NLS authors propose a dense zero-clipped reward that returns the makespan difference before and after each local search step induced by the action:

$$r = max(m_{t-1} - m_t, 0) \tag{6.7}$$

The details of the exact models used as teacher models are given in the next section which describes the dataset generation procedure.

Dataset Generation

In order to train the DTs in a supervised manner, a suitable labeled dataset must be created, as visualized in Fig. 6.7 under *Dataset Generation*. The dataset is generated based on 1,000 randomly generated JSSP problem instances per problem size in the Taillard benchmark dataset [22] and according to the same reported specifications (cf. Sect. 2.1.3). These instances are solved with the teacher NLS models in the environment provided in the code repository of Falkner et al. [42]. During the search process, the state-action-reward tuples, represented as (S, A, r) in Fig. 6.7, of trajectories consisting of 100 and 200 local search iterations are collected. 100 and

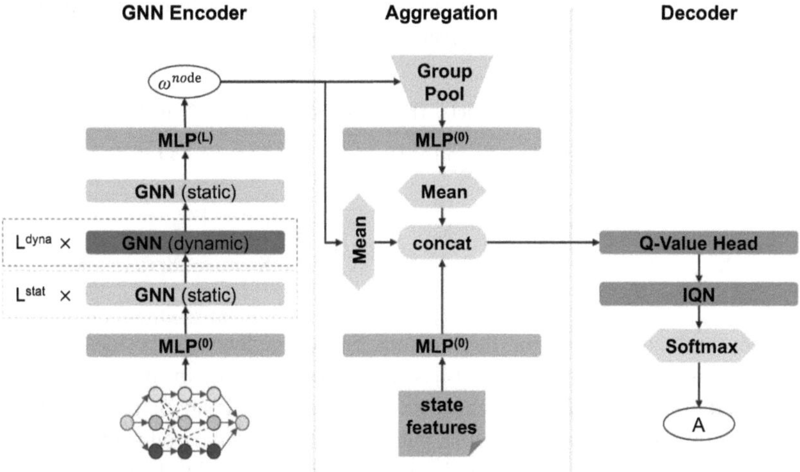

Fig. 6.9 NN architecture of the NLS approach (adapted from [42])

200 iteration steps are in the same range of the number of iteration steps investigated in the original NLS publication [42]. The saved trajectories form the preliminary raw labeled dataset.

The teacher models are pre-trained NLS checkpoint models that are available online in a repository provided by Falkner et al. [42] for each problem size. The idea behind choosing the pre-trained models is that the DTs can learn from the already successful strategies of trained NLS models. The paper reports that different action sets led to the best results for different problem sizes. For example, action set A_A showed the best performance for 15 × 15 problems and action set A_{ANP} for 30 × 20 problems [42]. With the goal of generating the dataset from trajectories taken by the best possible teacher models, the pre-trained models with the best corresponding action set per problem size are used. However, there are two exceptions, problem sizes 15 × 15 and 20 × 20. In these two cases, A_A was originally reported to perform best, but preliminary experiments conducted for the study in this subchapter showed that the NLS teacher models had learned to take the same action (1 = "*accept*") exclusively and the resulting DTs overfitted to this strategy, leading to the exact same strategy and test results. As a remedy, the next best action set, A_{AN}, was tried. This strategy worked well for 20 × 20, but the 15 × 15 A_{AN} teacher model had also learned to take exclusively the "accept" action, such that the third best teacher model, A_{ANP}, was finally chosen. An overview of the chosen action sets per problem size of the NLS teacher models is given in Table 6.4.

Table 6.4 Action sets of teacher models. Those used in this study are marked with an "X".

	15 × 15	20 × 15	20 × 20	30 × 15	30 × 20	50 × 13	50 × 20	100 × 20
A_A	–	–	–	–	–	–	–	–
A_{AN}	–	X	X	X	–	–	–	–
A_{ANP}	X	–	–	–	X	X	X	X

The third component for the training dataset generation, besides training instances and teacher models, is the NLS environment. Again, the environment is adapted from [42] and implements the corresponding action dynamics and evaluation routines. First, as in the original publication [42], the environment constructs the initial solution for any given training instance using the FDD/MWKR composite dispatching rule introduced by Sels et al. [279], which combines information on the flow due date and the remaining work of each job. As mentioned above, the teacher models and environments subsequently interact for 100 and 200 search steps, resulting in two generated raw datasets of state-action-reward tuples with 100,000 and 200,000 raw data points, respectively:

$$dataset_{raw} = \left[\underbrace{\{S_0, A_0, r_0\}^0}_{one\ raw\ datapoint}, \ldots, \{S_T, A_T, r_T\}^0, \ldots, \{S_T, A_T, r_T\}^i \right], \quad (6.8)$$

where $T \in [100, 200]$ are the number of search time steps on one training instance and $i \in [0, \ldots, 1000]$ is the number of training instances. 100 and 200 interactions were chosen because they were studied in the original publication, too. However, the original code does not provide returns-to-go. For DT training, the required returns-to-go are therefore calculated in an additional step by traversing each trajectory backwards and recursively adding the rewards, such that the returns-to-go R_t in each time step t become $R_t = \sum_{t-T}^{T} r_t$. The final datasets have the following format:

$$dataset_{final} = \left[\underbrace{\{S_0, A_0, R_0\}^0}_{one\ final\ datapoint}, \ldots, \{S_T, A_T, R_T\}^0, \ldots, \{S_T, A_T, R_T\}^i \right] \quad (6.9)$$

In summary, the major required modifications to the original NLS code base are a method to store the trajectory data and a method to transform the trajectory data into the DT format. The most important design choices lie in choosing the appropriate trained teacher models and the number of local search iterations those teachers should perform as demonstrations.

DT-Training

Each DT model is trained for 500 epochs through gradient descent along the categorical cross-entropy loss between the predicted action $a_{t,pred}$ and the label action a_t in the dataset, as shown in Fig. 6.7. Preliminary experiments with varying hyperparameters on the 15×15 JSSP indicate that a context length K of the DT between 50 and 200 leads to the best policy improvements. As a tradeoff between effectiveness and a larger required model for large K s, $K = 50$ is chosen in this study. The Adam optimizer [280] is used with cosine learning rate decay to smoothly reduce the overall learning rate with periodic increases back to the initial value of 0.06. A buffer is implemented that parses the dataset into the proper format for the DT, including the context of previous interactions and the current state and return-to-go. For iteration steps t that are smaller than the context length K, the remaining input tokens are zero-padded and zero-masked within the self-attention mechanism. The model is tested every 33 epochs on 30 JSSP instances of the respective problem size on which the model trains, generated with a different random seed than the training and evaluation instances. Whenever the achieved test makespan improves, the model weights are saved.

The proposed DT NN architecture combines elements of the original NLS model and a transformer architecture. The idea is to utilize the learned latent space representation generated by the aggregator component of the NLS model as meaningful state tokens. For the dataset generation, the original pre-trained checkpoint model weights are used, which are available online [281]. The transformer architecture is based on an implementation of the minGPT model by Karpathy [282], from which most of the hyperparameters have been adopted. These hyperparameters are listed in Table 6.5.

The transformer in the DT replaces the decoder component of the original NLS model to generate the action distribution in the output layer, as depicted in Fig. 6.10. The state embeddings, actions and rewards are all projected into the same embedding dimension of the tokens through linear layers. The resulting sequence of tokens is then fed into the first multi-head attention layer of the transformer. The combination of the two frameworks in the proposed way requires only little changes to both code bases. One conceptual modification is that, in contrast to the original implementation of minGPT [282] for textual data, the positional encoding is not created in relation to

Table 6.5 Hyperparameters and design choices of the transformer model in the DT [43]

Hyperparameter / Design Choice	Value / Name
number of layers	6
number of attention heads	8
embedding dimension	128
batch sizes	512
context length K	50
nonlinearity	gelu
dropout	0.1
Adam betas	(0.9, 0.95)
grad norm clip	1.0
weight decay	0.1
learning rate decay function	cosine decay

the input block size but the maximum trajectory length of the JSS solution generation corresponding to the number of iteration steps of the NLS.

DT-Testing
Finally, certain design choices need to be made for the testing phase of the DT. These choices also define how the DT is used during inference. The conceptual representation of the testing procedure is provided in Fig. 6.7. As described before, the original NLS DRL environment returns the clipped relative makespan improvement caused by the last action as reward signal, starting at time step zero with an initial solution for the given JSSP instance. During testing or inference with the DT instead of the DRL agent, this poses a challenge, because appropriate return-to-go values are needed as input, not rewards. Ideally, at time step zero, the return-to-go R_0 would be the difference between the makespan of the initial schedule m_{init} (obtained from the FDD/MWKR dispatching rule [42]) and the makespan of the optimal solution $m_{optimal}$.

However, $m_{optimal}$ is usually unknown prior to solving a problem instance. The alternative feasible proposition is to compute a lower bound on the makespan m_{lb} as in equation (12), defined as the maximal sum of those processing times p of operations O_m that require the same machine m:

$$m_{lb} = \max_{machine \in M} \sum_{o \in O_m} p(o) \tag{6.10}$$

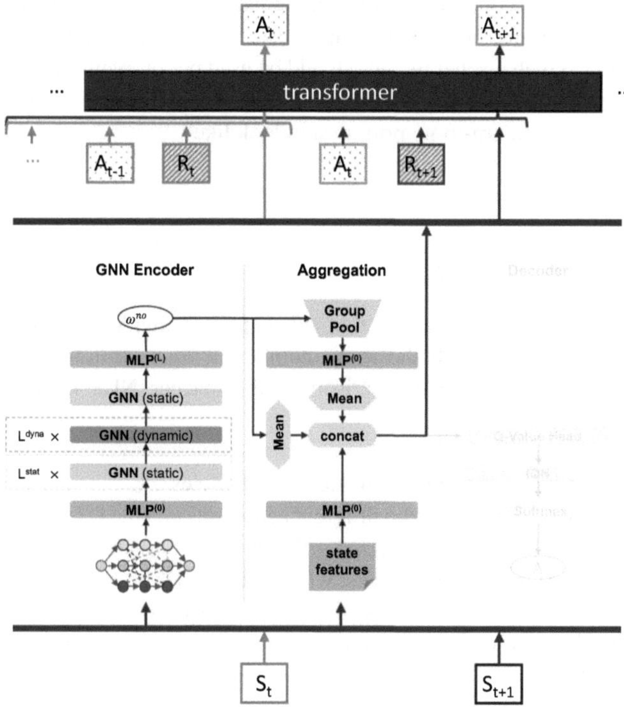

Fig. 6.10 Combination of the NLS model and the transformer implementation for the DT (parts adapted from [43])

The lower bound makespan m_{lb} essentially ignores any precedence constraints within jobs, such that $m_{lb} \leq m_{optimal}$. The initial return-to-go at step zero R_o is then computed by $R_o = m_{init} - m_{lb}$ and is reduced in each NLS iteration by the reward returned by the NLS environment. The crucial motivation for this procedure is that R_o is an optimistic approximation, forcing the DT to take the best possible solution to reach m_{lb} from m_{linit}. Additionally, as in the DT-training procedure, an intermediate buffer is implemented to store the historic trajectory data needed as DT input, since the original NLS environment would otherwise not return in each iteration step.

In summary, to arrive at the final method for dataset generation, DT-training and DT-testing, the existing NLS code base and model weights are combined with an existing implementation of a transformer architecture. For clarity, the major own design choices in this study beyond the combination of the two methods concern the following questions: how much training data is generated and from which trajectory length? Which teacher models should be used per problem size? What is an appropriate choice for the state token embedding form the trained models? What does an appropriate return-to-go prior design look like?

6.2.3 Results

The presentation of the results of this subchapter is divided into the numerical results on test data with the same number of local search iterations and a subsequent more in-depth comparison of the teacher NLS models and the corresponding DTs.

Numerical Results on the Quality of Learned Policies

The results of the NLS models with different action sets and the DTs on the Taillard benchmark dataset [22] are summarized in Table 6.6. DT_{100} and DT_{200} refer to the DT models that are trained on datasets, where the teacher NLS models performed 100 and 200 search iterations, respectively. Note that regardless of the training dataset, both DTs perform 200 search iterations during inference. For each model, the average makespan (m) achieved for each problem size and the average optimality gap (gap) are given. The results of the NLS models with action sets that were not used for the dataset generation, i.e., that are not used as teacher models, are grayed out. The best overall model for each problem size is printed in bold.

Apparently, the DTs can outperform their NLS teacher models in almost all problem sizes and in some cases by significant margins. Across all problem sizes, the DT_{100} leads to optimality gaps that are 1.11 %p, 1.23 %p and 1.15 %p better than NLS_{ANP}, NLS_{AN}, or NLS_{A} individually. Such improvements are non-trivial for the JSSP and fall outside of random noise artifacts. While DT_{200} and DT_{100} perform similarly well on average, each taken alone does not consistently outperform the NLS models on all problem sizes. In practice, this means that both variants should be tried to achieve the best final result for a given problem size.

To validate the observations on a larger set of problem instances, the results on 100 randomly generated JSSP instances are reported in Table 6.7. The results largely confirm the above observations on the smaller Taillard benchmark dataset. As before, the DT models outperform the NLS models on average. However, compared to the previous test results, they do so by smaller margins and not as consistently. These

Table 6.6 Results (makespans and optimality gaps) on the Taillard benchmark instances

		15 × 15	20 × 15	20 × 20	30 × 15	30 × 20	50 × 13	50 × 20	100 × 20	average
NLS_A	m	1324.0	1530.7	1804.0	2043.1	2267.0	3273.0	3163.0	5682.0	2635.85
	gap	7.74 %	12.16 %	11.54 %	14.33 %	19.42 %	18.00 %	11.22 %	5.89 %	12.54 %
NLS_{AN}	m	1336.0	1520.2	1806.0	2042.7	2270.0	3281.0	3161.0	5679.0	2636.99
	gap	8.72 %	11.39 %	11.67 %	14.31 %	19.57 %	18.29 %	11.15 %	5.84 %	12.62 %
NLS_{ANP}	m	1357.0	1578.1	1841.0	2033.9	**2261.0**	3068.5	3152.0	5673.0	2620.56
	gap	10.42 %	15.63 %	13.83 %	13.82 %	**19.10 %**	10.62 %	10.83 %	5.73 %	12.50 %
DT_{200}	m	1335.0	**1517.0**	1807.0	2039.3	2263	**3066.4**	**3145.0**	5681.0	2606.71
	gap	8.63 %	**11.15 %**	11.73 %	14.12 %	19.21 %	**10.55 %**	**10.59 %**	5.88 %	11.48 %
DT_{100}	m	**1323.0**	1529.3	**1802.0**	**2030.4**	2264.0	3071.7	3152.0	**5664.0**	**2604.55**
	gap	**7.66 %**	12.05 %	**11.42 %**	**13.62 %**	19.26 %	10.74 %	10.83 %	**5.56 %**	**11.39 %**
Opt.	m	1228.9	1364.8	1617.3	1787.0	1898.4	2773.8	2843.9	5365.7	2359.98

results and the comparison show that the quality of the generalizing policies of the DT_{100} and DT_{200} models varies between different specific problem instances.

Behavior Comparison between NLS teacher models and DTs

Next, the practicality and learned behavior of the DT models are examined in more detail and in comparison to their respective teacher models. Conceptually, the two main differences between the NLS teachers and DTs are as follows: First, the DT models can base their decisions on a context of up to 50 previously observed states, taken actions and received returns-to-go. This allows them to take advantage of the information about the influence of past decisions of their local search on a currently given problem instance. Indeed, in preliminary experiments, we observed performance improvements only when context lengths of 50 steps or more were used. On the downside, to take the larger context into account, a much larger NN architecture is used, which leads to increased inference times compared to the original NLS architectures. Second, a major difference from DRL models that aim to maximize the cumulative rewards is that DT models have a manually designed return-to-go prior that the models are trained to match, and which can be used to push them toward shorter makespans than those observed during training. These differences lead to the following three questions:

1) What are the practical implications of using the comparatively larger and slower DT models during inference with respect to performance?
2) Is there a correlation between the relatively better performance of DT models compared to their teacher models and a greater deviation in learned behavior?
3) Is there an optimal return-to-go to achieve the best performance?

The results in Table 6.6 and Table 6.7 indicate better individual decisions made by the DTs than their NLS teacher models, since the same number of search iterations leads to better makespans. However, the inference time of the DT is longer than that of the smaller NLS models on the same hardware.

The practical impact of longer inference times can be seen in Fig. 6.11. Plot a) on the left shows the mean makespan as a function of the number of search iterations of the teacher (NLS_{ANP}) and the DT model on our 15×15 JSSP instances. The gap between the models increases with the growing number of iterations, i.e., the DT model continues to improve the solution faster with more steps. Plot b) in Fig. 6.11 on the right shows the same achieved mean makespans in relation to the computational time required for the complete search on our hardware (Intel Core i9-9980HK CPU @ 2.40 GHz with 64 GB RAM and NVIDIA Quadro T2000 GPU). Since the teacher model performs at par with the DT for small numbers of iterations,

Table 6.7 Results (makespans) on 100 randomly generated test instances

	15 × 15	20 × 15	20 × 20	30 × 15	30 × 20	50 × 13	50 × 20	100 × 20	Avg.
NLSAN/NLSANP	1319.7	1499.4	**1736.8**	1972.2	**2178.2**	2974.9	3137.0	5694.7	2564.1
DT200	1310.0	**1496.4**	1739.5	**1972.1**	2181.0	**2974.7**	3138.1	5689.6	2562.7
DT100	**1301.3**	1499.2	1737.0	1972.6	2181.1	2975.3	**3136.9**	**5675.2**	**2559.8**

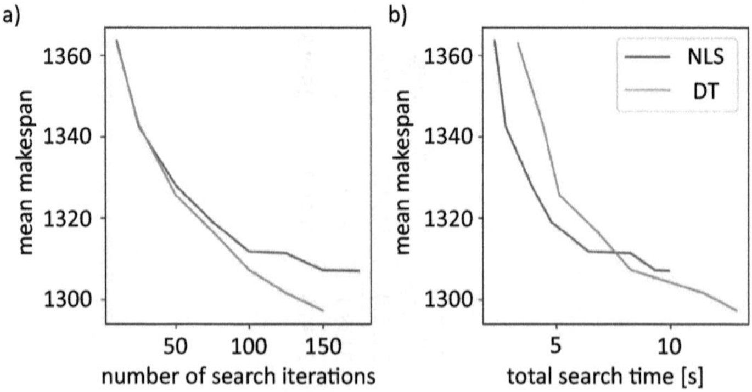

Fig. 6.11 Comparison of achieved average makespans by NLS and DT models with respect to the number of iterations (a) and the total search time in seconds (b)

but is faster, it outperforms the DT up to search times of about 7 seconds. However, since the teacher model converges to a worse average makespan, the DT achieves better makespans for all search times greater than seven seconds. Therefore, in practice, it is advantageous to use the DT when search times greater than seven seconds are acceptable. It is important to note that the point at which this tradeoff occurs can vary significantly depending on the final performance of the DT and the size of the problem instances. In some cases, such as the 20×20 JSSP, the DT may not outperform the teacher model.

To compare the learned strategies between teacher and DT models, the frequency with which available actions are chosen when solving the 100 generated instances are analyzed next. One may hypothesize that DT models achieving makespans similar to their teacher models may have simply learned to replicate the demonstrated behavior. The action frequencies of the 15×15 DT, representing the DT with the greatest relative improvement over its teacher model, are shown in Fig. 6.12a). The 20×20 and 50×20 DT models, as representatives with similar performance to their teacher models, are shown in Fig. 6.12b) and Fig. 6.12c), respectively. The available actions on the x-axis comprise the neighborhood operators CT, CET, ECET and CEI, and the perturb operator, each in combination with a reject or accept action (odd and even numbers). As expected from the performance difference, the action frequencies vary between the teacher and the DT model on 15×15 instances. While the frequency of taking the "CT + reject" action 0 is similar, the DT model has learned to never take the ECET actions 4 and 5, and strongly prioritizes the "CET + accept" action 3 in favor of fewer "CEI + not reject" actions 7. The action frequency distribution

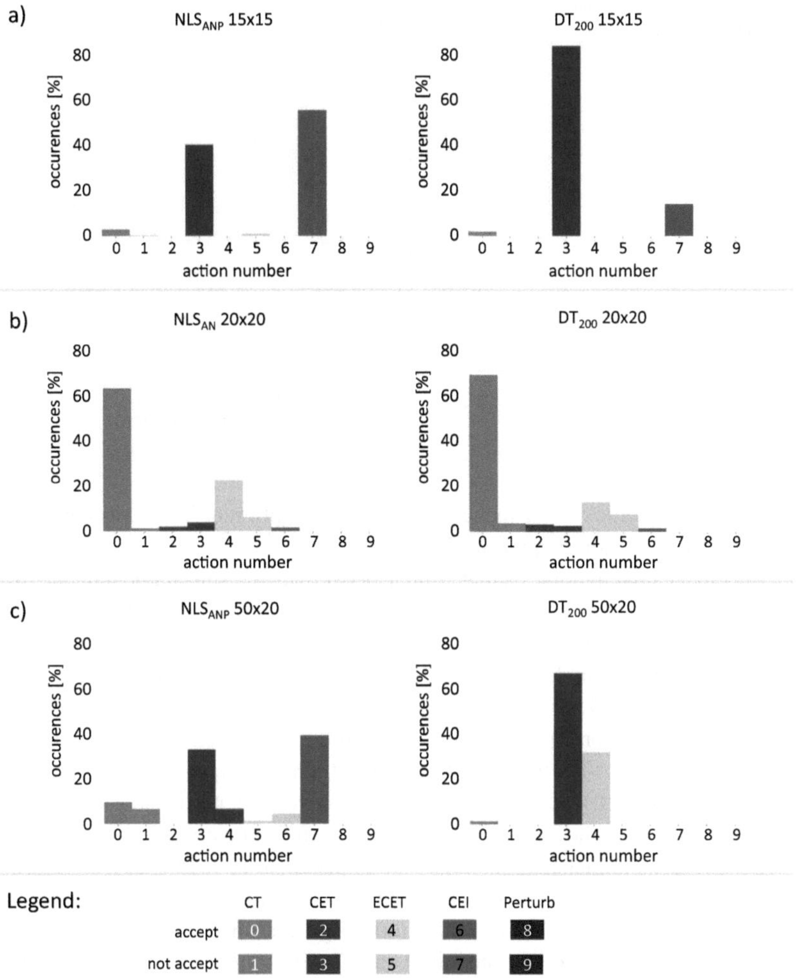

Fig. 6.12 Action frequencies of NLS and DT models

in 20×20 is, in comparison, much more similar between the teacher and the DT model. This, together with the similar makespans reported Table 6.7, suggests a very similar behavior of the two models. Contrarily, 50×20 presents itself as a

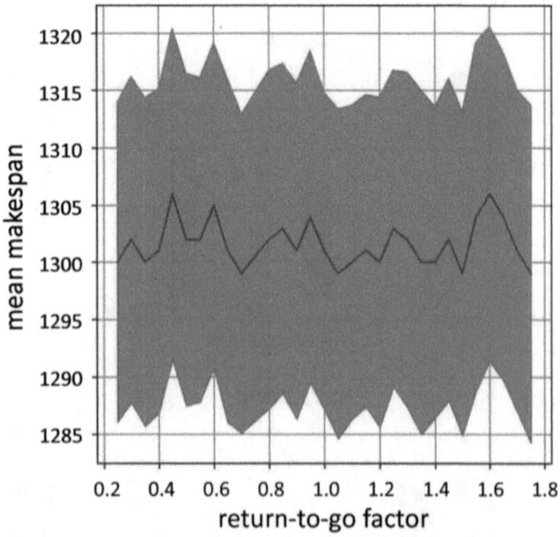

Fig. 6.13 Mean makespans of the DT on 100 15×15 JSSP instances over varying return-to-go factors

counterexample with similar makespans but a very different learned behavior, as expressed by the action frequencies in Fig. 6.12c).

This falsifies the hypothesis that similar performance results alone indicate that the DT models are merely copying the teacher's behavior. An interesting observation is that the perturbation operation actions 8 and 9 were not performed by either the teacher or the student models that include these actions (NLS_{ANP} and DT_{200} 15×15, NLS_{ANP} and DT_{200} 50×20). This is noteworthy because the ANP action sets, which contain these operations, performed better than the AN action sets. This behavior is an artifact of the NLS training, not the DT training. However, it serves as an indication that the DT learns to ignore actions that were not used by the teacher and is therefore limited in how much it can deviate from demonstrated strategies.

To analyze the influence of the return-to-go prior, the 15×15 DT_{100} on the 100 randomly generated instances is tested while multiplying the initial lower bound makespan m_{lb}, which was used as the return-to-go starting point during inference with return-to-go factors in the interval [0.05, 1.75] in 0.05 increments. The interval is chosen such that the resulting return-to-go priors cover the achieved makespans of the DT_{100} model. The results are shown in Fig. 6.13. There is no visible trend in the mean makespan across different return-to-go factors. Additionally, the 95 % confidence interval indicated by the shaded area, is much larger than the difference between the minimal and maximal values. The conclusion is that the return-to-go has no statistically significant influence on the resulting mean makespans. Interestingly, all obtained mean makespans are smaller than those of the teacher model. This means that the DT has learned a new superior strategy independent of the return-to-go.

6.2.4 Summary and Discussion

This subchapter introduced a method that deploys a DT to learn from local neighborhood search steps taken by a trained DRL when solving JSSP instances. The results of the study indicate that the trained DTs can significantly outperform their teacher models, and can even compensate for increased inference times that are due to the much larger NN architecture of the designed DT by better individual actions. In principle, the DT models have two advantages over their teacher models: the context of how previous search steps influenced the solution and a manual return-to-go prior that can push the DT towards better actions. Indeed, it could be shown that distinct behaviors are induced in DTs compared to their teacher models. However, the experimental results suggest that only the first advantage has an influence on the learned behavior, while the return-to-go prior has no significant impact.

RQ3 asks, how a trained DRL agent should be used, given a fixed computational budget during inference, to decrease the expected value of the minimal makespan of the generated solution. With respect to this research question, the presented study serves as an example that trained DRL agents can be successfully used as teacher algorithms to generate expert trajectories from which a student model can learn. If properly designed, this student model can then show superior performance in terms of decreasing the expected value of the minimal makespan with the same computational budget. In a broader sense, the presented study also serves as an example that building on top of the artifacts of other researchers' methods is worthwhile—a notion that is far from new in other research fields like computer vision, where it is common practice to use pre-trained models—instead of designing completely distinct solutions for closely related problems, as is still common practice in DRL-based scheduling.

Despite its contribution, the study serves more as proof of concept that a student model can outperform a converged DRL model than a proof that his method can be applied to any DRL model trained to solve the JSSP. Therefore, one limitation of the method is that it is likely only to be successful for metaheuristics where a context of historical actions provides meaningful information for how to proceed. In theory, learned construction heuristics do not fulfill this property condition. At the same time, at least in the current state of the art, learned construction heuristics are superior to learned metaheuristics on the JSSP. However, proponents of learned metaheuristics argue that they are more easily adaptable than learned construction heuristics when moving away from the JSSP to more real-world scheduling problems [283]. From this perspective, research on this class of algorithms is highly relevant. Outside the scope of this thesis but in future work, it would be interesting to expand the study to include more learned metaheuristics. Promising candidates with good performance on JSSP-related scheduling problems have been developed by Reijnen et al. [283], where a DRL agent learns to control parameters of an evolutionary algorithm, and by Chen and Tian [199], who introduced local rewriting, where an agent iteratively points to local regions in the solution graph of a problem instance and chooses a rewriting strategy for that region.

In addition to the design choices for the teacher model, the choice of the DT as a representative of offline RL algorithms has its limitations. This is especially evident in light of the mentioned lack of influence of the return-to-go prior, being one of the distinguishing features of the DT, on the performance of trained DTs. Other newer algorithms, which were briefly discussed in the related work in Sect. 6.2.1, may even outperform the DT in terms of inference speed or quality of actions.

Generally, in terms of superior net performance regarding inference speed and the quality of actions, this subchapter introduced a method by which a larger and slower NN learns to make higher-quality decisions than a converged DRL teacher model. As described this can lead to net positive results, if the resulting policy is of sufficiently higher quality. The opposite approach is also conceivable. As outlined in the related work section of this subchapter, Sect. 6.2.1, knowledge distillation aims at training a much smaller and faster NN with only slightly worse decision-making capabilities. If the ratio of the quality of decisions to inference time is sufficiently small, one can also expect a net positive effect from more neighborhood search steps in the same inference time. This possibility is promising for future work and has the potential to provide further insights regarding RQ3.

From the JSSP to Real-World Applications

7

The previous Chaps. 4–6 introduced novel techniques for improving DRL-based solution methods to the standard JSSP, focusing on achieving shorter makespans within a given time. One of the quintessential prerequisites before practitioners will consider deploying a DRL-based solution is achieving competitive objective values compared to alternative solutions. However, real-world machine scheduling use cases are likely to diverge from standard JSSPs addressed thus far. In practice, it may be infeasible to generate one plan offline and follow it strictly. Adjustments will likely be necessary to account for changing or new environmental conditions. This chapter explores the implications of transitioning from the academic JSSP setting to the real-world industrial application settings for DRL-based machine scheduling solutions. Thereby, it addresses the fourth, open research question of this thesis (RQ4). Specifically, it provides preliminary answers to the question of which additional challenges arise when transitioning from the abstract JSSP to real-world production planning scenarios, and how these challenges may be addressed.

In a first step, the scope of a real-world scheduling system as it is considered in this chapter is introduced, including involved elements and their interactions, to provide a framework for further discussions. Figure 7.1 illustrates these different elements as well as their interactions. They comprise

- the environment, that encapsulates the scheduling problem, i.e., the machine setup, additional constraints and influencing factors of the machine scheduling problem, as well as the evaluation criteria,
- the DRL agent that interacts with the environment through perceived environmental states (observations) that form the basis for actions of the DRL agent, that are subsequently rewarded,

C. Waubert de Puiseau, *Advancing Training and Inference Methods for Reinforcement Learning-Based Job Shop Scheduling*,
https://doi.org/10.1007/978-3-658-51305-4_7

- and a user who interacts with the scheduling system by potentially entering additional data and preferences into the environment and using the scheduling system to create a schedule either autonomously or with agent assistance.

Together, the depicted elements and their interactions outline the structure of a prospective comprehensive DRL-based machine scheduling solution and how it may function in practice. Note that the user is the primary novel element when transitioning from creating solutions in an academic setting and making a usable scheduling system. Beyond the user, the central element subject to change is the environment. To asses prospective changes to the environment, it is useful to revisit the vision behind DRL-based JSSP, as it was described in Section 2.3.2. This vision describes moving toward problem settings that make decisions based on an information-rich observation, aiming to optimize relevant real-world objectives—or in other words, it describes the vision of end-to-end DRL-based machine scheduling. Note that the depiction of the vision inside the environment component is a slightly shortened version of Fig. 2.8. More changing dimensions can thus be found there.

The scope of this chapter does not allow for going into full detail on the changes to every involved element and their possible interaction patterns. However, the necessary adjustments to solution methods can be categorically different depending on which element is subject to change. Consequently, to lay the groundwork for future studies, the following SubSection 7.1 provides only few examples of consequential changes to each element that were found in the literature. The examples are an attempt to allow the reader to build an intuition for the implications of transitioning towards applications that are closer to real-world scenarios. The subchapter also offers preliminary ideas for dealing with resulting challenges. These ideas are not exhaustive, but they provide guidance for potential solution paths.

In addition to the more general overview of changes to individual elements in the system, SubSection 7.2 aims to provide a deep dive into one additional evaluation criterion for scheduling systems that gains particular significance when transitioning from academic trials to practical application: reliability. Reliability is crucial in industrial settings, since system failures, unexpected outputs and inconsistent behavior carry high costs [284]. Reliability as a system characteristic encompasses all the aforementioned elements (user, environment, and DRL agent), making it a challenging yet fascinating real-world requirement. However, given that little work has been done on real-world problems (cf. Chap. 3), much groundwork for structured future research is missing, including a clearly defined terminology, metrics, as well as adjacent research directions

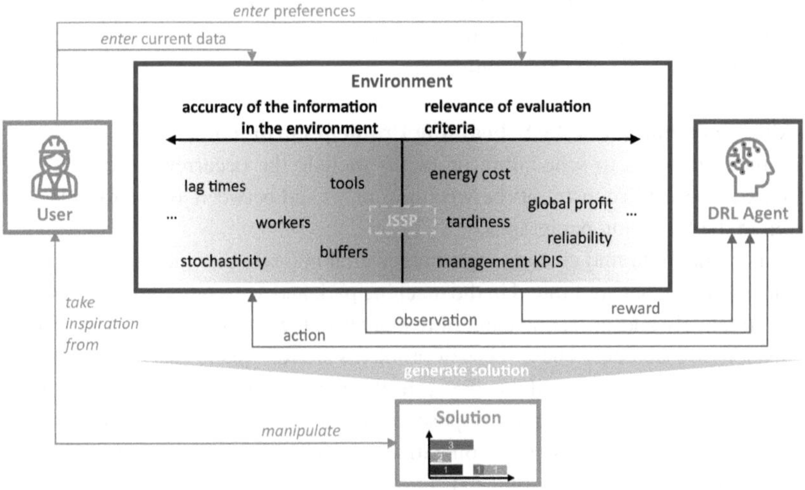

Fig. 7.1 Interactions in a real-world DRL-based scheduling system

for increasing the reliability of DRL-based machine scheduling problems. Therefore, SubSection 7.2 is dedicated to building this groundwork in two sections. First, Section 7.2.1 defines reliability criteria for DRL-based machine scheduling systems and suggests metrics for their quantification based on a literature review. Subsequently, Section 7.2.2 discusses preliminary ideas for methods to increase the reliability of a complete DRL-based machine scheduling system. Finally, SubSection 7.3 summarizes the results and contributions of this chapter and discusses limitations.

7.1 Additional Challenges and Promising Remedies

First, exemplary changes regarding additionally considered elements within the environment are examined. These changes correspond to adaptions that model the characteristics of certain real-world scheduling setting more accurately, schematically depicted as moving left from the central "JSSP" in Fig. 7.1. Individual adjustments, such as the logical incorporation of tools, different machine layouts, finite buffers, supply chains, and human workers, may necessitate fundamental adjustments to the design decisions that were discussed in Section 2.3.1. The following selected examples from the literature that are representative of frequently occurring additional challenges. The goal of discussing them is to gain an understanding for implications of certain changes and derive common methodological approaches to create effective solutions.

Adding Information to the Scheduling Problem Formulation

Common additions to scheduling problems include the occurrence of additional resources as well as constraints between resources and between jobs. Examples are given in the following paragraphs.

It is common to find changes regarding available resources, such as introducing additional tools that are limited in the machine park and are required for operations. This effectively adds a second class of resources to the problem that can be seen as equivalent to machines. These types of problems are referred to as a multi-resource constrained scheduling problems [285]. In these problems, compared to the JSSP, the scheduling of one operation imposes constraints that affect a greater number of other operations, because these operations cannot be scheduled in parallel while their required resources are blocked. From a user's perspective, creating a solution for such a machine scheduling problem becomes cognitively more challenging, because one has to "keep in mind" this greater number of immediate effects of scheduling a single operation when planning ahead. From a computational perspective, however, the implications are different. For example, if two additional tools are considered, and if each operation requires exactly one specific one of these tools, then the number of possible solutions remains the same. In other words, the solution space size remains the same. Thus, formally, the problem does not become more complex. The only difference is that on average, solutions will have greater makespans, because fewer operations can be scheduled in parallel. This example shows that the implications of an additional constraint can be counterintuitive and affect different components of the scheduling system differently. From a DRL-based solution design perspective, the learning problem of a multi-resource constrained problem is arguably more difficult than the JSSP. This is primarily because the size of an observation space that provides full information on the new problem increases at least linearly with the number of additional resources. This, in turn, requires some adjustments to the NN architecture. Apart from these two changes, there is no direct motivation to make further problem-specific changes. The action space, reward design, DRL algorithm, and other design choices could remain conceptually similar to those of a functional design for the JSSP.

Adding constraints to resources can have similar effects. For example, limiting the capacity of buffers for jobs before and after machines [105, 107], logically only requires adjusting the observation to capture additional information. For example, this could include the current occupation level of buffers. However, exactly because such additions logically only require small changes, it is difficult to determine what to change, if a DRL agent fails to converge during training on such an enhanced scheduling problem. Consequently, and somewhat counterintuitively, this poses a major challenge for engineering a novel solution, because many possible design decisions (cf. Section 3.2.3) and hyperparameters could be changed.

Of course, there are also changes to the scheduling problem that clearly require adjustments beyond the observation and NN architecture. Additional constraints within jobs often have this property. This is especially true if certain sequences of operations result in infeasible schedules. A common example in practice is that operations can be subject to strict time constraints between each other. These constraints may define the time between the start and completion of one operation and the start and completion of another operation. This problem is referred to as a scheduling problem with time lags [286]. Time lags are best explained by example. Consider operation A that involves the application of an adhesive to a part. Due to chemical properties of the adhesive, there is a fixed time interval between a minimal and maximal dry within which another part must be applied (operation B) for the adhesive to function properly. In such a case, it is likely that not only the observation, but also the action logic must be redesigned significantly, so that the agent is biased towards satisfying this constraint, or so that a whole critical block consisting of operations A and B is scheduled at once. It is also likely that not satisfy the constraint must be punished through a dedicated term in the reward.

Moving Towards Real-World Evaluation Criteria

Clearly, changes to the problem characteristics have far-reaching engineering consequences. Next, exemplary changes to the evaluation criteria, in other words the objective of the optimization problem, are considered. At first glance, addressing divergences from the makespan only influences the reward function. However, additional features in the observation space of the agent are also necessary. For instance, optimizing the tardiness requires adding a due date for each job to the observation and a reward term that penalizes the difference between the due date and the completion time. Similarly, when trying to optimize the energy cost under periodically varying costs depending on the time of the day, one must establish an information basis via the observation that includes an absolute time reference. This is necessary so that the agent can know the absolute time it schedules operations. This information is not necessary in the JSSP. In such a case, one would also likely make changes to the action logic to facilitate scheduling operations more densely around cheaper electricity prices. For example, one could apply an automated shifting towards such time periods instead of applying left-shifting (see Section 3.2.3, "action logics").

An example of a more visionary evaluation criterion is the contribution to optimize the global profit of a factory, or even a whole company, through machine scheduling. Naturally, optimizing a goal from a higher-level perspective without losing sight of the details of the machine scheduling problem requires expanding the observation space. Another consequence is that the time window that must be observed by the agent grows. Machine scheduling is a downstream task from events such as changes in demand and the supply chain. Concurrently, it is an upstream step from tasks like final quality checks, packaging, delivery, or other logistic tasks before an order is completed. The longer relevant time horizon before a job enters the shop floor requires the agent to observe and make predictions based on earlier events that will influence the downstream scheduling task. Vice versa, the agent should observe and consider the state and performance of tasks that follow later when a job leaves the shop floor to adjust the production rate accordingly. Thus, the challenges include a drastically increased observation space size and corresponding NN architecture, more time delayed rewards because the evaluation of successful schedules takes place further downstream, and the increased number of environment element interdependencies that must be implicitly learned. A possible remedy for these challenges could include the division of necessary predictions and actions into multiple subtasks handled by multiple models in a hierarchical DRL manner [287]. Other options include model-based DRL approaches that will make it easier to evaluate the long-term impact of predictions and actions [288], and other techniques that address the longer time delays between actions and rewards, such as Evolution Strategies (cf. Section 2.2) or synthetic returns [289].

From the above examples of additional elements in real-world problems, it is apparent that, although there are certain promising ideas for addressing them, one can only speculate how future solutions will be composed exactly. Since most of the work on more application-focused problem settings is done by researchers in the computer science, operations research, and manufacturing domain rather than to researchers in AI (cf. Section 3.2.1), progress will likely continue to be made in incremental steps rather than through one fundamentally disruptive approach. Most likely, many seemingly incremental steps in adding real-world elements will require both, a suitable combination of several nuanced changes to existing design decisions, but also adoptions of entirely novel concepts developed in the AI domain. These concepts include novel DRL-algorithms, disruptive NN architectures like the transformer [251], hierarchical learning strategies [287], and mixture-of-expert methods [290].

Implications for Involved Elements of DRL-based Scheduling Systems

The challenges discussed so far mostly concern the DRL algorithm design along the necessary design decisions introduced in Section 2.3.1. What they have in common is that they require environments in which more computations occur to update the state between state transitions. Apart from learning-problem related challenges from the DRL agent's perspective, this also has two practical implications for engineers training and implementing these algorithms, because more detailed simulations are required. First, creating a comprehensive simulation is more time-consuming and difficult. Second, simulation times will increase considerably. Regarding the increased manual effort for creating simulations, it is likely that generative large language models will increase the productivity and capabilities of simulation engineers in the short term, and in the long term be capable of automatically translating real-world constraints into functional code for complex simulations. The first steps into this direction have already been taken for simulations in other domains such as climate simulations [291]. Furthermore, creating Python-based simulations is becoming easier, as significant progress is made in Python-Code generation [292]. This is relevant since most DRL software frameworks are currently written in Python, facilitating seamless integrations for environments written in Python [61, 293, 294]. Concurrently, commercial players like Siemens are considering incorporating AI-assisted simulation engineering into their products [295]. The second implication, slower simulations, can also pose major challenges. They can quickly become a limiting bottleneck during all phases of development and training, and also during the inference phase of the trained DRL-based solution. Developments that could mitigate this negative impact include GPU-accelerated discrete event simulations [296], parallelized environments [293], and NN-based surrogate models that replace certain otherwise sequential simulation steps [297, 298]. However, the challenge of slower simulations when more details are included persists.

Another likely consequence of increasing the complexity of simulations is an increasing gap between simulations and reality (sim2real gap). This is because inaccurate assumptions about system dynamics could potentially accumulate. Therefore, strategies learned in simulations and deployed in reality could be ineffective. However, recent results on other tasks support the fact that learning with imperfect simulation dynamics can still work during application or only require minimal further training in the real environment [299]. Furthermore, potential remedies such as domain randomization in simulations [300] exist to mitigate the sim2real gap.

Finally, humans are involved not only in the engineering phase, but also as users in the usage phase, as depicted in Fig. 7.1. At least in the short term, it is likely that DRL-based scheduling systems will serve as assistance systems rather than autonomous schedulers. Hence, the goal of DRL-based machine scheduling should be to optimally support a human planner (hereafter referred to as the "user") to improve schedule quality and reduce the manual planning effort. In the simplest case, all information is automatically integrated into a digital shadow that serves as a scheduling environment. A schedule is then created through interactions between a trained DRL agent and this environment. The resulting schedule can then be used either as a fixed schedule to follow in reality, or as inspiration for manually creating a schedule. A third alternative, which lies in between, is the assisted generation of a schedule. In an assisted generation method that follows the constructive heuristic solution approach (cf. Section 2.1.3), the user can overwrite the DRL agent's action while still following the same action logic as the agent. Alternatively, the user could manipulate a completed schedule. Thereby, the user acts more similarly to a local neighborhood search operator (cf. Section 6.2.2). Of course, a user may also indirectly explore different solutions by manipulating job priorities in the problem description, changing constraints such as available machines, or changing shift plans. The introduction of a user also raises questions about the user interface design and the acceptance of a DRL-aided solution method in general. However, such considerations exceed the scope of this chapter.

In conclusion, on the basis of a framework of different elements involved in a practical machine scheduling system, this subchapter has explored examples for the multifaceted challenges that arise when applying DRL-based machine learning in real-world settings. These challenges, if viewed from the perspective of existing solutions to DRL-based JSS, require different adjustments and extensions to logically account for them (such as additional information in the observation space for additional resources). Individual adjustments may influence each other when they are combined to account for several deviations from the JSSP at once. It therefore is advisable to address any deviations step by step, trying to limit them to adjustments that are logically required (instead of fundamentally changing the NN architecture) to make incremental but steady progress.

7.2 Reliability of DRL-based Machine Scheduling

When transitioning from research and development in simulation to real-world usage, the demand for the reliability of the used scheduling system becomes more pronounced. Reliability is generally considered a positive characteristic. Yet, what reliability is exactly in the context of DRL-based machine scheduling often remains vague. The nonfiction book *Reliability Engineering* by Birolini provides a general definition of reliability as "the probability that the [...] [system] is able to perform as required for a given time interval" [284]. Therefore, reliability relates to the proper functioning of the entire scheduling system. According to this definition, all components of the scheduling system and all design decisions must be evaluated to quantify the systems reliability. Consequently, suggesting intuitive methods to increase overall reliability is not straightforward.

Birolini's [284] definition of reliability is very broad, leaving much room for interpretation regarding how one defines the scope of the "system", and what "performing as required" means. Regarding the scope, it is useful to distinguish between the reliability of the technological scheduling system (encompassing the information technology (IT) and the soft- and hardware infrastructure), the reliability of the user, and the reliability of the complete DRL method (comprising the environment and DRL agent).

An analysis of the technical reliability of the IT infrastructure (related to functional safety [301]) and the reliability of the user behavior are beyond the considerations of this chapter. In order to answer the general question of how reliably DRL can be used as a method for machine scheduling problems, a complete requirement list for the method and the types of problems it addresses would need to be defined. This also lies outside the scope of this chapter. Instead, the following sections cover a narrower scope, focusing on the reliability of scheduling strategies, rescheduling schemes, and of single schedules. First, suitable and quantifiable definitions of reliability within this scope are discussed in Section 7.2.1), before suggesting promising methods to increase the reliability accordingly in Section 7.2.2.

This subchapter contains information and ideas that have been published as a separate journal article [62]. However, it has been updated with additional thoughts that have since evolved, as well as with new candidate methods for increasing the reliability that have since been introduced in the literature.

7.2.1 Definitions of Reliability in Machine Scheduling and DRL

Given that DRL-based machine scheduling lies at the intersection of machine scheduling and DRL, it is helpful to distinguish between these two research fields when searching for respective definitions of reliability. This allows the identification of common ground, and definitions and metrics from the optimization literature that can inspire methods adaptable for DRL-based solutions.

Reliability Definitions in Classical Approaches to Machine Scheduling
From the perspective of classical optimization techniques, as opposed to DRL-based techniques, Leusin et al. [302] define a reliable dynamic scheduling system as one which fulfills three criteria. Namely, such a system

1) anticipates common disturbances in the production scenario,
2) can be adjusted to fit new circumstances with minimal change to the original plan,
3) and quickly reacts to such new circumstances by suggesting the best adjustments.

This definition relates to stochastic disturbances in the machine scheduling scenario. The impact of such disturbances and how they can be anticipated is described in an example by Pinedo [11]. Figure 7.2 depicts this example, in which three jobs must be sequentially scheduled on a single machine. There are no precedence constraints between jobs. Each job j is characterized by a unitless processing time t, a unitless due date d measured from the starting time of the problem for its completion time ct, and a unitless weighting value w for how much a job is worth if finished on time. If finished late, the value decreases proportionally to the tardiness. Hence, the objective is to minimize the weighted tardiness, $\sum_{j \in J} w_j * max[0, ct_j - d_j]$. . In a deterministic environment without disruptions, the optimal job sequence is $(1, 2, 3)$. In this sequence, all jobs are finished on time, resulting in an objective value of zero. However, in this example, there are random machine downtimes at the beginning of the process, causing initial delays. If there is an initial delay of ten time units before any operations can begin, the sequence (1, 2, 3) leads to a weighted tardiness of $10 * 1 + 8 * 100 + 6 * 100 = 1410$. . However, the originally suboptimal sequence (2, 3, 1) leads to a weighted tardiness of only $30 * 1 + 0 * 100 + 0 * 100 = 30$ when faced with the same disruption. The function of the weighted tardiness over time is shown in Fig. 7.2 on the left. According to Leusin's criterion 1), a system suggesting sequence (2, 3, 1) can be evaluated as more reliable, since it anticipates a disturbance more effectively.

Reliability criterion 2) goes in a similar direction, but instead of creating a reliable schedule in anticipation of random events, it assumes that the original plan can be slightly altered to account for the new circumstance without changing the entire plan. Intuitively, this reliability criterion is important, if workers receive a shift plan according to a production schedule. In such a case and if a disruption causes the shift plan to change significantly, it will cause logistical complications, frustration, and distrust in the scheduling system. Criterion 3) evaluates how quickly a system

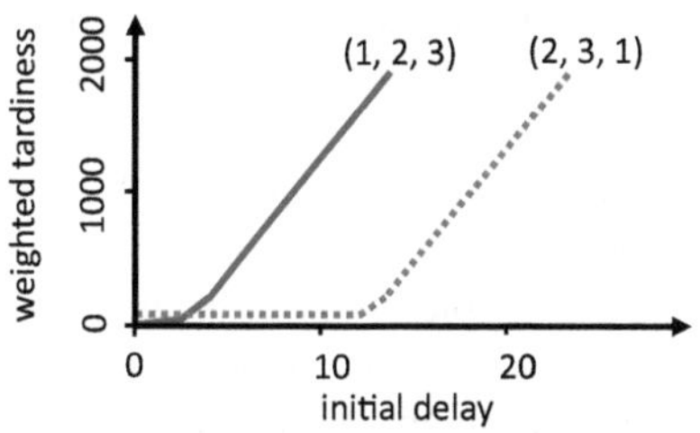

jobs j	1	2	3
duration t	10	10	10
due date d	10	22	34
weighting w	1	100	100

Fig. 7.2 Example for the reliability of two schedules under uncertainty

can react to new circumstances. This property is fundamentally tied to the implementation and integration into the IT infrastructure. It is therefore excluded from further discussions.

Examining the literature reveals synonymous terms for criteria similar to Leusin's [302] first criterion. Specifically, the terms *robustness* and *stability* appear in the context of executing a production process according to a schedule [303, 304]. Robustness generally describes that the performance metrics of a schedule are only minimally affected by a disruption [303]. Stability describes the property of a schedule, that disruptions have little impact on the timing of discrete events in the schedule [304]. In the above example, job sequence (1, 2, 3) is less robust than (2, 3, 1). However, they are equally stable, since the start of processing times will be delayed by 30 time units regardless of the order of jobs. Note that in other scenarios, the shift of events may be directly related to the performance metric.

Common metrics for quantifying robustness and stability use the expected production process under a given schedule as a basis for comparison with how the process actually unfolds, either in reality or in one or several simulated runs within stochastically parameterized simulations. Examples of such metrics from the optimization domain are listed in the following, because they help to define the concepts more concretely. To quantify robustness, the differences in expected objective values are often measured. For example, Rahmani and Heydari, as well as Shen et al. measure the differences in resulting makespans [305, 306] or the total workload of machines [306]. Analogously, several researchers calculate the sum of the differences between the expected and realized completion times of all operations to quantify the stability of a schedule [304, 306, 307]. Another option to quantify robustness is through the difference between the expected objective value and the worst realized objective value within multiple stochastic simulations. For example, the maximum makespan [163, 302, 308] and maximum tardiness [304, 309, 310] have been analyzed in the past. Another and very common group of metrics originating from risk-measures in finance [311] are the value at risk (VaR), defined as the worst quantile of the objective value, or the average/conditional value at risk (AVaR/CVaR), defined as the average of all values within a worst quantile [312]. Alternatively, changes to any objectives can also be divided by the duration of the causing disturbance to measure the sensitivity, or changes to objectives can be multiplied by the probability of a causing disruption to incorporate the likelihood as in a risk measurement [11]. Sotskov et al. [313] took a different approach by defining the so-called stability radius as the maximum change to the most sensitive processing time of a scheduled operation for which the schedule remains optimal.

There also exist reliability-related metrics of a schedule applicable before it has been executed in reality or in simulation. These include *flex*, defined as the number of pairwise jobs that are unrelated by precedence constraints, and *fluidity*, an estimate for the probability that localized changes to the schedule can be absorbed instead of propagating through it [314]. Unfortunately, these metrics are only defined for a so-called partial order schedule and are therefore not universally applicable [314]. Pinedo suggests that, when the bottleneck machine of a system has a high utilization, small deviations cannot be compensated easily, defining this property as a counterindication for reliability [11]. However, this metric requires the guaranteed identification of a bottleneck machine, which may be difficult or impossible in certain cases. Shen et al. [306] consider the presence of idle times between operations to be a characteristic that increases reliability due to their ability to absorb local disturbances. Unfortunately, this metric is highly negatively correlated with a short makespan. In summary, there is till now no universally applicable metric for evaluating the reliability of a single schedule prior to execution.

There are comparatively few metrics for reliability criterion 2), which addresses the ability to react to disturbances with minimal changes. This is because a single metric for this criterion requires more context for interpretation. Inherently, criterion 2) depends not only on the original schedule, but also on the rescheduling routine. According to Pinedo [11], rescheduling routines can be divided into three categories: "right-shifting rescheduling", which effectively puts the process on hold for the duration of the disruption before continuing as planned; "partial rescheduling", which makes adjustments to the original schedule only where it was affected; and "complete rescheduling", which treats the state after disruption as an independent, new problem. The "partial rescheduling" routine most directly considers the desirable minimal change expressed in criterion 2). However, measured reliability in this regard also depends on several factors. One such factor is the timing of the disruption, since a disruption towards the end of a schedule tends to create fewer necessary changes. Another factor is the deliberate choice of how the trade-off between an optimal new schedule according to the original problem objective and minimal adjustments to the old schedule is weighted in the rescheduling algorithm. To quantify the changes to the original schedule, Kouvelis and Yu [315] calculate the sum of all deviations of positions occupied by operations before and after rescheduling. Bean at al. [316] calculate the number of reassigned jobs. While these metrics can be useful, as mentioned before, they must be interpreted in the context of the entire rescheduling algorithm.

Reliability Definitions in DRL

Four areas of research appear in the context of reliability in the DRL literature: robustness, stability, safety, and performance consistency. Importantly, these terms are defined differently than in the classical optimization literature. Stability, for example, generally refers to the successful, reproducible, and smooth learning ability of an algorithm during training. This kind of stability is also sometimes, albeit inconsistently, referred to as robustness [317, 318]. As previously mentioned, this type of stability or reliability in learning behavior is not the focus of this chapter and is therefore disregarded. Descriptions and metrics for the remaining three areas are discussed in the following.

In DRL, robustness commonly refers to a limited sensitivity of the performance of a trained policy to stochastic variances, such as noise in the observation [319], disturbances to action signals the policy suggests [320], and stochastic transitions of the environment [321–323]. Unlike reliability definitions in the classical optimization domain, this definition evaluates the online scheduling performance of a policy or strategy rather than a set schedule. Instead of evaluating a fixed schedule, the agent is tested in multiple runs in the stochastic environment, and its reliability is quantified by measuring the standard deviation or the CVaR of the reward [321, 324].

Safety, as another notion of reliability, is defined as the satisfaction of constraints in the action space (cf. action validity in Section 3.2.3) or in the state space during training and testing. It is of high relevance in physical environments, where constrained state spaces can indicate positions or locations that an agent, e.g., a robot or autonomous car, should avoid to prevent collisions. It is therefore common that this type of reliability in DRL is also described as risk-sensitive, risk-adverse, or cautions behavior [325, 326]. Remaining within constraints is also important in scheduling. For example, extending due dates may incur penalties, or violating time lags between operations may render parts unusable. When there is a cost or penalty for not satisfying a constraint, popular metrics to quantify the reliability are the average or standard deviation of the cost [327, 328]. When constraint violations are binary, i.e., they are either satisfied or not, the number of constraint violations [11] or the percentage of tests ending in a predefined catastrophe can also be used to evaluate safety [329].

Lastly, a consistent performance is also perceived as a more reliable performance [328]. While DRL optimizes for expected returns and thereby towards achieving good average performance during training, one may also be interested in the deviations from the average performance. Inconsistent performance can have different root causes. One cause may be the trained agent itself, if it is only successful in a subset of the tasks given to it during evaluation. Another cause can be the agent's inability to compensate for non-deterministic transition behavior, which leads to larger performance deviations than desired. Note that if, as in the latter case, performance fluctuations are caused by stochasticity, the definition of inconsistent performance aligns with that of robustness. It can be difficult to attribute performance inconsistencies to their root cause. Suitable metrics are, similarly to those for robustness, the standard deviation of achieved returns, their inter-quantile-range, their CVaR, and the number of times the agent passes a minimal performance threshold [318].

Reliability in DRL-based Machine Scheduling

From the above definitions and metrics from the different research fields, a list of desirable properties and metrics regarding the reliability of DRL-based machine scheduling can be derived. For the definitions, it is helpful to adopt the differentiation from the optimization domain into three categories: the reliability of a single created schedule, the reliability of a trained policy, and the reliability in terms of the ability to reschedule well. Each of these can be evaluated with respect to different sub-criteria of reliability, the definitions of which are merged from those in the optimization and the DRL domains. These sub-criteria are: *safety*, from here on defined as adherence to constraints with or without stochasticity in the problem setting; *stability*, defined as minimal sensitivity of the structure of the schedule to noise in the observation and action space, or environment transitions; *robustness*, defined as minimal sensitivity of the achieved objective value with respect to such noise; *performance consistency*, defined as consistent performance when faced with different problem instances.

Accordingly, the proposal for how to evaluate the **reliability of a single schedule** is the following: as observed in the classical optimization domain, evaluating the reliability of a single schedule without access to a simulation is difficult. In terms of safety, one can test whether all constraints are satisfied and measure the distance to constraint borders. Measuring the robustness of a single schedule without simulation is infeasible. Since a single schedule comprises one problem instance, the consistent performance sub-criterion is not applicable. If there is a known bottleneck, the utilization of the bottleneck in the schedule indicates low robustness and stability, as do small idle times in general. However, these measures are rather weak and unspecific to DRL as a method for generating any individual schedule. The picture changes when a simulation is available to test the schedule thoroughly, which is usually the case because the simulation is a requirement for DRL training. There are three different possibilities for testing a schedule under stochastic conditions, depending on the definition of what defines a solution: the order of actions an agent took, the resulting order of operations on machines, or the order of starting times of all operations. Accordingly, these three possibilities are:

1) Saving and then deploying the action sequence of the agent, i.e., the order of prioritizations used to generate the original schedule.
2) Trying to keep the order of operations on each machine constant despite changing environmental circumstances.
3) Adhering to the order of starting times of operations in the original schedule.

For any of these possibilities, the realistic stochastic parameters in the simulation must be defined, such as stochastic processing times or machine failures, and then perform multiple simulations. Safety can then be measured according to the proposed metrics from the DRL literature. Similarly, robustness in terms of consistent performance can be measured using statistical metrics such as the standard deviation, interquartile range, or CVaRs. Since these evaluations are performed without querying the DRL agent again, the impact of noise in the observation space, action space, and transition probability can only be evaluated indirectly via the performance rather than by comparing single policy network outputs. This also means that the same simulated runs can be statistically evaluated with respect to the different criteria: robustness by evaluating the achieved objective values; stability by evaluating structural changes to the schedule, as described in the above paragraphs on reliability measures in the classical optimization literature.

The evaluation of the **reliability of a trained policy** can be conducted analogously. The prerequisite is a parameterizable simulation to evaluate a policy before execution in the real system. However, these evaluations can also be performed retrospectively after using the policy in the real system. The same metrics used for single schedule evaluation can then be applied to measure safety, robustness, and stability. The only difference lies in the fact that multiple instances will be evaluated, which introduces more noise in the metrics. Therefore, a large enough number of simulations are required to obtain statistically significant results. To quantify performance consistency, the policy should be evaluated on a pre-defined set of varying problem instances that capture the essence of the reliability criterion of interest. Additionally, since access to the policy is granted, one can conduct a dedicated analysis of the sensitivity of the policy to noise in the observation, measured through the Kullback-Leibler divergence between the outputs with and without noise. Importantly, this is only an indirect measure of the impact of stochastic noise on the scheduling performance. The impact of noise on performance can be measured more directly through statistical metrics of the achieved objective values according to the previously defined metrics for criterion 1) in the classical optimization literature.

A different approach is necessary for the evaluation of the **ability to reschedule well**. As mentioned before, this ability depends on both the original schedule and the rescheduling strategy in response to stochastic events. At the system level, detecting the necessity to reschedule also plays an important role. Furthermore, it is debatable whether a scheduling agent and a rescheduling agent will be the same module in a scheduling system, since their required abilities differ. For evaluation purposes, it is therefore most practical to examine the whole system end-to-end and compare schedules before and after rescheduling. In this case, the reliability metrics corresponding to criterion 2) from the above section on reliability in classical optimization can be used. These metrics are the sum of necessary operation shifts and the sum of the caused deviations of starting times, which relate to the stability of the rescheduling routine. To quantify safety and robustness in rescheduling, one can take inspiration from the metrics in their respective DRL sections, but regarding differences between constraint violations and objective values before and after rescheduling, rather than of differences between several stochastic simulation runs.

Table 7.1 lists the proposed reliability metrics grouped by safety, stability, robustness, and performance consistency. Note that as described above, although most metrics are universally applicable for evaluating a single schedule, a trained policy, and a rescheduling algorithm, the respective way in which the experiments are conducted varies considerably. Also, a single schedule cannot be evaluated for consistent performance.

Having defined the criteria for evaluating reliability in DRL-based scheduling, in the next step, the 143 articles in the literature corpus generated in Chap. 3 are revisited to analyze whether the proposed notions of reliability are actively sought in existing studies. This analysis helps to identify the overlaps between the proposed definitions and metrics and common practice in the state of the art, as well as potential research gaps. For stability and robustness, which are tied to stochasticity, only the subset of 60 papers that include stochastic elements in their environment were analyzed. Of these, only 20 consider the reliability of their respective proposed scheduling system. To include safety, which is not directly linked to stochasticity in the environment, a separate keyword search for the term "safety" was conducted and filtered to obtain studies that explicitly measure safety. This yielded only one publication. Regarding performance consistency, many studies analyzed the performance on different problem sizes and reported the standard deviation. However, none reported any other statistical values, which is why this criterion for reliability can be said to be only indirectly and not actively represented in the literature so far. Since further analysis would not be valuable, performance consistency is excluded from the remainder of this section.

Table 7.1 Overview of proposed reliability metrics

	Reliability Metrics
Safety	• % of tests with constraint violations • cumulative constraint violation value • average and variance of constraint violation values
Stability	• bottleneck utilization rate • deviation of task positions • number of re-ordered jobs • flex • fluidity • presence of idle times
Robustness (OV = objective value)	• bottleneck utilization rate • statistical measures* of OVs • stability radius • difference in OV divided by causing stochastic disruption • difference in OV multiplied by the probability of the disruption
Performance Consistency (OV = objective value)	• statistical measures* of OVs

* statistical measures: average, standard deviation, interquartile range, minimum, maximum, VaR/CVaR

Table 7.2 provides an overview of how the found publications can be sorted into the different reliability criteria (except for performance consistency) defined here for DRL-based machine scheduling. As can be seen, the coverage of reliability considerations in the literature is sparse. None of the articles specifically analyze the reliability of single schedules, indicating a wide research gap. The ability to reschedule well was also only analyzed in a single publication, in which the authors measured the difference between agent-based rescheduling and continuing with a fixed rule-based strategy [167].

The publication by Altenmüller et al. [107] is the only one to report a safety-related metric, in this case, the number of time constraint violations when testing on unseen data. While more publications regarding safety exist, they often only report an average reward in which a penalty term penalizes constraint violations, which is not a definitive reliability metric [93]. On the other hand, the study by Zhou et al. [106], is the only one that measures the stability of a trained policy by reporting the average change of processing times resulting from stochastic machine failures. Most reliability-related metrics were reported regarding robustness under stochasticity. However, all studies in this category only report simple statistics on the objective values, such the standard deviation or minimum and maximum values, as a byproduct of reporting their generalization performance. Wang et al. [168] are the only ones to analyze the smaller standard deviation specifically as a sign of robustness.

The infrequent use of reliability metrics is not surprising, given that reliability is a concept that only becomes more relevant when transitioning into the practical use of a DRL-based solution. Until now, no such cases have been reported, at least not in the literature analyzed in Chap. 3. However, as research progresses, it is likely that more and more concepts that have already been established in the more mature optimization domain will become relevant.

Table 7.2 Publications considering different notions of reliability

	Single Schedule	Trained Policy	Rescheduling
Safety	–	[107]	–
Stability	–	[106]	–
Robustness	–	[59, 96, 105–107, 113, 114, 116, 117, 123, 157, 163–165, 168, 175, 186, 189, 192]	[167]

7.2.2 Methods to Increase the Reliability of DRL-based Machine Scheduling

As could be seen in the previous section, reporting explicit measurements of reliability is not yet common practice in DRL-based machine scheduling studies. Therefore, it is not surprising that only minimal work exists on methods to increase the reliability of DRL agents for machine scheduling. To provide a starting point, this section outlines further ideas for such methods, based on two assumptions: that the path towards considering reliability is inevitable; and that if reliability metrics are consistently reported and seem promising, the application of DRL-based machine scheduling systems can be accelerated. The ideas in this section draw inspiration in part from techniques described in the optimization and DRL literature but are also offer own novel suggestions.

Several conceivable methods for increasing reliability equally apply to all three levels, a single schedule, a policy, and a rescheduling scheme, because a better policy improves all three. Therefore, the following discussion begins with conceptual methods for improving a scheduling policy with respect to reliability, before moving on to rescheduling.

A universal and straightforward strategy to increase stability and robustness, which has already been applied in many studies, is to incorporate relevant stochastic elements into training. This exposes the DRL agent to disturbances and allows it to implicitly learn a more reliable strategy. However, this is not an explicit reliability technique and already standard procedure in many cases. For instance, agents in scheduling problems with random machine breakdowns are typically trained in environments that model random machine breakdowns according to a predefined probability distribution [104, 114, 189]. In the scheduling domain, actively introducing more noise than would be expected in the deployment phase is less common and could be beneficial. This could be done manually through techniques such as domain randomization [300] or through a learned adversarial agent that is trained with a reward opposite to the reliability criteria. Unfortunately, stochasticity can make training more challenging [44]. If training under stochasticity leads to unstable learning behavior, curricula that gradually increase the severity of stochastic disturbances over the course of training could be an effective countermeasure to facilitate successful training.

Further methods consider more than just the introduction of stochasticity in the environment during training. The other design decisions beyond the environment, such as the reward, observation space, and action space, also have the potential to incentivize more reliable policies. For instance, any of the introduced reliability metrics that apply to a single schedule can be used to design a sparse reward signal at the end of each episode, complementing the performance-related reward design. In the observation space, any information about expected stochastic disruptions, such as predictions of machine breakdowns, could also help the agent to learn not to rely heavily on machines that are likely to break down soon. In the action space, one could try to incorporate stochastic forward simulations before making a decision to evaluate the riskiest trajectories and mask out the corresponding, less reliable next actions.

Notably, methods to increase the reliability of a machine scheduling system can also be applied during the usage phase of the agent. If the available time and hardware allow, it could be useful to explore multiple possible trajectories after each possible next action in simulation, to get a statistical analysis of the implications of this action with respect to the objective, but also with respect to reliability metrics. For such simulations, MCTS is a suitable method for balancing visit-counts and node evaluations, as has been done for machine scheduling by Li et al. [93] in the safety context. However, since sampling strategies are parallelizable, not just with a stochastic policy but also in a stochastic environment, they could also be an effective and perhaps more efficient option with a similar effect. In this context, one should also consider deploying several different trained agents with varying levels of the desired reliability criterion in parallel, in order to generate more diverse candidate schedules. For this purpose, it would also be an option to train one model with multiple heads that represent different reliability levels, as has been successfully done to create and leverage different strategies in the PolyNet approach [330] for vehicle routing problems.

In practice, it is unlikely that online scheduling will be used. Online scheduling would imply that an agent is queried just in time to make decisions when they are needed. It is more likely that a schedule will be created as a plan to be followed. No matter how well the policy used for the plan generation is trained to anticipate disturbances, mitigate the effects of disturbances, and avoid unsafe states, there will always be cases in which rescheduling is unavoidable. One reason is that there will likely be a tradeoff between the expected strategic objective (e.g., the makespan) and the reliability of the schedule. Therefore, a certain level of risk will likely always be accepted to reach favorable objective values. A second reason is that not all corner cases can be accounted for in the strategy. Third, any realistic scheduling situation will change at some point in time when new jobs are introduced into the system. This makes rescheduling a necessity in practice.

Any rescheduling scheme involves three components: an original schedule, a component that identifies situations in which a rescheduling action should be taken, and a component that carries out the rescheduling action itself. Between the states in which everything is happening according to the original plan, and the state in which following the original plan becomes infeasible, there is a phase in which it is non-trivial to decide whether to trigger a rescheduling routine or not. Hence, a monitoring system must be implemented to determine the exact point at which rescheduling is beneficial. This system could involve a separate DRL agent that collaborates with the scheduling agent at a lower decision frequency, sharing a common reward term.

The choice of a rescheduling routine that schedules unfinished operations in the system to account for new environment conditions depends largely on this routine's objective. If the objective is the same as that of the original schedule, then one may also use the same agent and inference method that was used to create the original schedule. However, if part of the goal is to reschedule with minimal changes to the original schedule, then a separate method should be used, potentially in the form of a separate DRL agent. This separate DRL agent optimizes a multi-objective function composed of the objective of the original plan and an objective that involves the desired reliability metrics for rescheduling (e.g., limiting unwanted changes to the original plan). A rich body of literature on multi-objective optimization through DRL exists from which inspiration can be taken [331]. There are also specific approaches for generating a Pareto set of solutions with DRL that weigh multiple objectives in scheduling problems [146, 332].

Of course, the complete solution composed of a system for creating the original plan, a monitoring system for the identification of a rescheduling point, and a system for rescheduling, does not have to involve three DRL agents. Depending on the problem setting, it is likely that a manual, rule-based, or other alternative approach may be able to fulfill one of these roles. In any case, the involved DRL agents will have to adapt to the respective other two components and must be at least partially rewarded according to the joint scheduling performance by means of a common element in their reward signal. It is also advisable to fix the strategies of two components while training or changing the third to avoid oscillations in learning behavior.

7.3 Summary and Discussion

This chapter addressed the additional challenges that arise when transitioning from the JSSP to applying DRL-based machine scheduling solutions to real-world problems. A framework was provided to attribute these additionally arising challenges to different components, including users, simulations and DRL methods. The chapter also suggested preliminary ideas to address these challenges. Particular attention was given to the notion of the reliability of DRL-based machine scheduling systems, as this is an essential requirement for their deployment in practice. Based on literature from the classical optimization domain and DRL, definitions of the reliability of a DRL-based machine scheduling system were derived and quantitative metrics were collected to measure the reliability of a schedule, a policy, and a rescheduling method. Finally, ideas for increasing the reliability of DRL-based solutions were outlined, providing a foundation for the necessary future work to make DRL applicable to real-world scheduling problems.

In doing so, this chapter has made several contributions regarding RQ4 of this thesis, namely, which additional challenges arise when moving beyond the JSSP in an academic setting and how these challenges may be addressed. As these are open research questions, the contributions in this chapter are of conceptual nature, rather than explicitly suggesting and thoroughly evaluating particular methods for particular use cases. Nevertheless, the results provide valuable orientation points for future work.

One contribution lies in the suggested framework (cf. Fig. 7.1) for attributing elements that appear in a real-world scenario to different components of a prospective scheduling system. This framework enables future contributions to real-world systems to be communicated to the research community and practitioners in a more structured and standardized way than before. The given examples of emerging challenges resulting from changes to the problem setup, evaluation criteria, and growing simulations help to foster this proposed structure. Furthermore, examples of how different challenges can be approached from different perspectives, along with concrete pointers toward promising remedies, serve as guidance for fellow researchers in this field.

Another significant contribution is the aggregation of reliability criteria and metrics applicable to the DRL-based machine scheduling domain. While the terms robustness, stability, and reliability are currently used interchangeably to describe different characteristics in different contexts, these terms and corresponding definitions can now be used to describe and measure different notions or reliability more precisely. Finally, the ideas provided for increasing reliability, inspired by established concepts and current research trends, offer initial approaches that could lead to significant progress towards the real-world applicability of DRL in machine scheduling.

Despite these considerable contributions, the content of this chapter is not without limitations. Rather than providing a comprehensive list of possible changes, the chapter merely described single examples of changes within components of the DRL-based machine scheduling framework. The same is true of the corresponding methods used to mitigate additional challenges resulting from these changes. While this certainly limits the value of the explanations for researchers addressing a change that has not been explicitly covered, it is unfortunately impossible to achieve complete coverage due to the large number of possibilities for and combinations of changes. Nevertheless, the demonstrated approach of identifying implications of certain changes to necessary design choices remains a useful change-agnostic process.

Another limitation is the chapter merely lists suitable reliability metrics without ranking their usefulness. It is unlikely that every metric will be analyzed in depth in every future publication that considers reliability. Therefore, the metrics provided in Table 7.1 are intentionally left unfiltered and should be interpreted as suggestions. The hope is that this list is useful for initiating a structure conversation about how informative certain metrics are for specific problem categories and expressed requirements in practice. The structure could enable researchers to converge on a common metric more quickly in this discussion.

There is much uncertainty surrounding the difficulty of the learning problem for the DRL agent. Concurrently, given the rapid development in recent years, there is much uncertainty about the introduction of disruptive novel AI solutions in the near future. These uncertainties, coupled with the fact that even seemingly small changes to the problem can require significant changes to the overall DRL design in order to work, render it difficult to make suggestions about what a future scheduling system will look like. Therefore, while the suggested multi-agent approach comprising three separate agents for creating a schedule, monitoring and deciding a rescheduling point, and then rescheduling, is an intriguing concept, it will require significant development to work successfully.

This chapter did not discuss alternatives derived from existing classical optimization techniques for scheduling and rescheduling, because its intention was to outline a vision for DRL-based methods specifically. Therefore, the suggested visionary approach is strongly biased towards DRL and only works under the assumption that significant progress will be made in this area in the near future. In the meantime, however, established concepts from the classical optimization domain and their combination with DRL are certainly not irrelevant in future work—on the contrary numerical optimization approaches are still favorable, especially in safety-related problems, because they provide optimality and constraint satisfaction guarantees. Future work towards real-world applications should therefore carefully evaluate the strengths and weaknesses of different approaches to the different subtasks of a comprehensive scheduling and rescheduling solution. The important individual strengths and weaknesses of different approaches, such as optimality guarantees, the ability to handle stochastic environments, and inference times, must be considered and combined effectively to achieve an optimal system.

8 Critical Discussion and Outlook

This chapter critically reflects on the contributions of this thesis, evaluating their potential impact on the development of future DRL-based machine scheduling systems, but also their limitations. Notably, each chapter and subchapter that directly addresses a research question (Chaps. 4–7, and Subchapters 6.1 and 6.2) already features a dedicated discussion of its respective contributions and limitations. These in-depth analyses are only briefly summarized here, and readers seeking more details are referred to the corresponding sections. Instead, the emphasis of this chapter is on evaluating the addressed research questions and contributions in the broader context of advancing DRL-based machine scheduling. Specifically, the discussion in this chapter considers the following guiding questions:

- Which contributions does this thesis make to achieve the goal suggested by the title of this thesis, i.e., for advancing training and inference methods for DRL-based JSS?
- How do the studies and insights of this thesis contribute to achieving the larger vision of end-to-end DRL-based solutions for machine scheduling problems?
- How transferable are the gained insights and introduced methods to DRL-based solutions for other scheduling problems and their respective methods?
- What are promising next steps in this domain?

These questions are first discussed in terms of the individual research questions of this thesis, before discussing overarching limitations and avenues for future research.

C. Waubert de Puiseau, *Advancing Training and Inference Methods for Reinforcement Learning-Based Job Shop Scheduling*,
https://doi.org/10.1007/978-3-658-51305-4_8

Research Question One

The first research question (RQ1) asked how domain knowledge can be advantageously integrated into a DRL-based JSS system. The main contributions in this regard are twofold: a method to integrate bottleneck information inspired by the SBH into the observation space, and a comparative study of two DRL-based methods to evaluate the effect of providing this additional information on the training speed and final performance of the trained agent. However, since the study showed indifferent results along these dimensions and adding bottleneck information increased the required computation time significantly, the suggested approach was not advantageous but rather serves as a negative example for the inclusion of domain knowledge. Therefore, this particular method should not be applied in future studies to achieve better performance. Nevertheless, the study is still valuable. It shows that the integration of domain knowledge does not trivially lead to better solutions, even if a particular element of domain knowledge is crucial in a related successful solution approach. There are several possible explanations for this. In the presented case, the SBH-based bottleneck information can perhaps be trivially derived by DRL-agents from other features and is therefore redundant. Alternatively, the information is too difficult to interpret which renders it useless, or the information is not useful within the action logic of the DRL-agent. A fact supporting the latter argument is that whereas the DRL-agents construct a solution operation-by-operation, the SBH constructs a solution machine-by-machine. Hence, the bottleneck information may not be useful in the operation-by-operation approach. In either case, the fit between the DRL-agents design and the provided domain knowledge seems to not have been productive. Accordingly, an insight for future studies on domain knowledge integration is that it could be valuable to adapt ideas and elements of approaches that are conceptually more closely related. For example, the used bottleneck information could have been more useful for a DRL-agent that also generates solution machine by machine, like the SBH. However, this hypothesis remains to be shown in future work.

Looking into the future and at the potential to transition from the JSSP to real-world scheduling problems, the contribution of the specific suggested method is very limited, because the SBH is a JSSP-specific heuristic as is its notion of bottlenecks. One could even argue that this negative example may discourage future attempts from making use of domain knowledge. Furthermore, one could criticize the research question itself, arguing that relying on domain knowledge is not aligned with the vision of end-to-end DRL given that domain knowledge is needed for engineering more sophisticated features. However, there are counterarguments to be made in favor of the relevance of RQ1 in the future. One argument is that domain knowledge of experts on a particular real scheduling system is likely more specialized, abundant, and more important for solving that problem effectively. Additionally, domain knowledge is not limited to descriptive characteristics of a production scenario, such as a bottleneck resource. Further ideas may be explored in the future. For example, domain knowledge also encompasses procedural expertise in how schedules are created for a particular system, which can inspire effective novel action logics. In such a case, domain knowledge could be integrated without hindering end-to-end DRL but rather make it more effective. Lastly, the integration of experts in the design phase of an AI-assisted system can increase trust and acceptance of the system. Current approaches do not explicitly serve as assistance systems but as standalone solutions. As agents are increasingly capable of making good scheduling decisions, a possible next step is the structured derivation of requirements of the capabilities of agents that are integrated into the backend of an AI-assisted scheduling system. Thereby, target inference times, useful action logics, and target qualities of reached objective values could be derived that ensure future advancements are made in the right direction.

Hence, while the specific method explored in this thesis may not advance the state of the art much directly, integrating interdisciplinary domain knowledge has great potential to make DRL-based scheduling systems more effective and widely adopted in practice.

Research Question Two

The second research question (RQ2) addressed the design of efficient and effective training strategies for DRL agents on the JSSP. This thesis contributed a novel curriculum learning (CL) strategy grounded in a novel metric for the difficulty of particular problem instances derived from the most-tasks-remaining priority dispatching rule. The introduced CL strategy not only led to better results than before, but the curriculum also had a consistent effect on the learning behavior. This motivates future work in this direction to understand DRL-based JSS better, transfer the results to larger problem sizes and more recent DRL studies, and find overall applicable superior training strategies. The transfer of the CL strategy to other DRL-based methods both for the JSSP and other machine scheduling problems is conceptually feasible, but problem-specific adjustments to the difficulty metric may be necessary. The proposed method deployed one of the simplest and most effective priority dispatching rules and ordered the instances by the achieved objective value. Therefore, a natural starting point when transferring the method to other problem types is to identify a priority dispatching rule that works well on the particular problem, and sort problem instances by the respective objective value of interest.

The practical relevance of the contribution for future applications is limited by the uncertainty regarding the extent of performance improvement that can be realized, and if it is worth the additional required implementations and experiments during the design phase. The additional efforts stand in contrast to one of the evaluation criteria for scheduling algorithms introduced in Sect. 2.1.3, namely, the required expertise and implementation effort. However, CL could play an even more important role when transitioning to real-world problems. One argument supporting this claim is that in many real-world problems, the amount of available historical data for agent training is likely insufficient. This necessitates the synthetic creation of additional data. During this process, the questions of how to synthesize this data and in what order to use it during training arise naturally, making the adoption of CL a logical next step rather than generating additional overhead.

Furthermore, as discussed in Chap. 7, real-world problems shift towards multi-objective scheduling as well as multi-agent approaches for joint scheduling and rescheduling systems. Such problems inherently lend themselves to curriculum-based learning, where subtasks are learned sequentially (e.g., first optimizing one objective and then the next), and where strategies must adapt to the behavior of other agents while these remain fixed. Hence, RQ2 and the role of CL in answering it will likely be an important building block for developing more effective DRL-based machine scheduling systems in the future. The next step in this direction is to study the potential to increase the observed positive effects of CL even further by generating even harder training instances. Pushing this strategy to its limits could provide further insights into what exactly causes the observed effects on agent training. With respect to real-world problem instances, it could be valuable to study the effects of training on problem sizes that are significantly smaller or larger than the actual production setting faced in reality. While this implies more engineering effort for generating additional simulations and synthetic training data, varying the problem size over the course of training has shown consistent positive results in related studies on academic problems.

Research Question Three

Contributions with respect to the third research question (RQ3), addressing the efficient usage of trained agents, were made in two separate studies. The first study introduced adaptive action sampling as a novel method for fine-tuning the exploration-exploitation behavior of a trained agent, enabling more effective use of the available computational time during inference. The method showed very promising results and is easily transferable to DRL-based methods, whether they are for the JSSP, other machine scheduling problems, or even other combinatorial optimization problems. The easy transferability makes adaptive action sampling a particularly impactful contribution. On the other hand, the underlying concept of the deliberate use of an exponent (also referred to as temperature parameter) during inference is already standard practice in LLMs, limiting the novelty of the method.

Action sampling combines parallelized search with agent-based heuristic decisions. In real-world scenarios, it may be more challenging to find schedules that satisfy all constraints or achieve a desirable tradeoff between multiple objectives. Furthermore, as outlined in Chap. 7, creating multiple solutions in simulation is essential for assessing reliability. Future work should investigate the possibility of adjusting the sampling behavior to effectively influence certain reliability aspects, e.g., finding safe (constraint-satisfying) solutions, balancing a tradeoff between two objectives, or increasing the performance consistency through a favorable exploration-exploitation tradeoff during inference.

The second contribution regarding RQ3 is the usage of a DT to learn from and improve upon the policy of a trained DRL agent. On a broader scale, the method contributes an example for the successful use of such an agent by means of supervised learning without having to go through expensive simulations needed for DRL training. It is the first study in the machine scheduling domain to explore this path. In contrast to the prior work where DRL agents followed construction heuristic approaches, the DT was applied to a trained DRL agent steering a local search heuristic. The rationale behind this was that the DT utilizes information from previous steps and that these are more informative in local search than in constructive approaches. However, this limits the applicability to other constructive approaches that remain dominant, particularly in the literature on the standard JSSP.

This limitation may be less critical when moving towards real-world problems, where metaheuristics are often the preferred option due to their implementation simplicity and state-of-the-art performance. As described in Sect. 6.2.4, improving decisions and parameterizations in such metaheuristics with DRL is a promising research direction. Moreover, since metaheuristics iteratively improve upon existing schedules through targeted manipulations, they are especially well suited for rescheduling tasks, where an existing (partial) schedule must be adapted in response to changes. While constructive approaches provide the agent with more control over the solution generation process, which is more aligned with the vision of end-to-end DRL, future work could explore novel action logics of DRL agents to manipulate an existing schedule. This way, the respective strengths of metaheuristics and DL would be effectively merged. In this context, the introduced DT methodology may prove valuable for refining and improving these novel solutions.

Research Question Four

The fourth research question (RQ4) investigated the implications for current and future methods of transitioning from academic to real-world scenarios in the DRL-based machine scheduling domain. The main contributions in this regard include: a structured framework of involved components that a real-world system encompasses; preliminary ideas of how to methodologically address certain changes to these components; and a detailed consideration of the notion of reliability. The distinct aspects of reliability in the DRL-based machine scheduling context were defined and methods proposed to increase reliability in these aspects. Since RQ4 explicitly addresses real-world systems, the contributions are naturally relevant for machine scheduling beyond the JSSP and future research in this direction. A detailed discussion of the specific contributions and limitations is provided in Sect. 7.3.

Further General Contributions and Limitations

Beyond the insights of the individual studies in this thesis, certain contributions and limitations apply equally to multiple studies. One common characteristic is that while superior performance could be demonstrated through CL, action sampling, and the application of the DT, the improvements are rather incremental both in terms of the novelty of the proposed approach and in terms of the achieved optimality gaps. All three methods added single novel elements to existing solutions. However, this modularity has a substantial advantage. When changing or adding single novel elements, any performance improvements are directly attributable to this element. This contrasts with many publications in the related work which introduce several novel elements within entirely self-engineered overall DRL-designs, making it difficult to attribute potential improvements to individual elements within their designs.

Nevertheless, it is also valuable to explore radically different approaches that diverge further from established constructive approaches. For example, a recent suggestion in the literature on DRL-based vehicle routing problems is to use a DRL agent for deconstruction of a solution in combination with a local heuristic reconstruction. This suggestion has shown promising results [333]. Previous studies in that domain also concentrated on constructive logics. Similarly visionary approaches for scheduling problems could be promising directions for future work. When trying to conceptually transfer this approach to the scheduling domain, two major ideas emerge that have been underexplored in previous research. First, the idea that the combination of DRL-based decisions for particularly impactful decisions (identifying a point for solution deconstruction) and simple heuristics for less impactful routines (reconstruct a solution) leverages the particular strength of each solution paradigm in the tradeoff between the quality-of-solution and computational time.

Second, the idea that it could be easier to identify areas for improvement in an existing solution than to build a new solution from scratch. Hence, building on top of these ideas, a proposal for future work is to combine slower and faster DRL-agents for more difficult and easier tasks. For example, a slower agent could generate a good initial solution, a second slower agent identifies an area for improvement in the schedule or solution graph, and a third and faster agent generates multiple solution alternatives. Such a novel design requires rethinking how the environment works, calibrating the learning speed of the different involved agents, shared reward designs, and other fundamental design decisions. Yet, as shown for vehicle routing problems, such an endeavor could lead to drastically better results in the end.

A limitation of this thesis in light of the latest developments is that it focused specifically on DRL-based machine scheduling solutions. Yet, recent evidence suggests that self-labelling or self-improvement approaches for training DL models for constructive scheduling methods achieve performance superior to DRL algorithms [56, 57]. In future work, one could consider these alternatives to strict DRL approaches as well. Notably, all techniques introduced in this thesis are equally applicable to these alternative learning paradigms. Thus, evaluating their potential to improve the performance in non-DRL contexts presents another interesting avenue for future research.

Finally, as DRL-based solution methods are becoming more competitive to existing solutions, future research should integrate personal preferences of users of scheduling systems to increase the usefulness and acceptance of the solutions in production. Users may have personal or technical preferences that they either do not want to or cannot formulate explicitly. It would be an interesting avenue for future research to adopt elements of reinforcement learning from human feedback (RLHF) [334], which has gained traction as a subsequent step to supervised pre-training in large language models as a means of integrating human preferences. Applied to scheduling, such a method would involve users that rank different schedules by their preference or users that adjust DRL-generated schedules according to their preferences. The corresponding data captures preferences implicitly. After an initial DRL-agent training, agents could then be fine-tuned on this additionally created data. This method could be used to iteratively personalize and improve a scheduling agent in a scheduling system that is used in production over time.

Summary 9

This dissertation thesis addressed deep reinforcement learning (DRL) as a modern, data-driven solution approach for machine scheduling problems. The goal was to explore different paths for advancing existing training and inference methods of DRL agents that lead to scheduling systems which can be trained more efficiently and generate better schedules in short computation times. The job shop scheduling problem (JSSP) was selected as the primary object of study, as it not only lies at the foundation of many real-world machine scheduling problems but also remains a central benchmark problem in academia. The availability of standard benchmark datasets and solution methods for the JSSP facilitated the evaluation and comparison of proposed algorithmic advances with the state of the art. A comprehensive literature review and four research questions provided the framework for the investigations in this thesis.

The literature review covered 143 publications on DRL-based machine scheduling. It highlighted the growing scientific interest in this area, the interdisciplinary nature of the involved research community, and the dominance of studies focusing on abstracted scheduling problems such as the JSSP. Publications on job shop problems (JSPs), a subclass of machine scheduling problems that includes the JSSP, were analyzed in greater detail to identify recent contributions and algorithm design trends. This review revealed several research gaps and underexplored opportunities for improvement, which were addressed by the four research questions.

The first identified research gap was the limited integration of domain knowledge from operations research (OR) and manufacturing into DRL-based methods. This motivated the question of how such knowledge could be integrated advantageously in DRL-methods to improve learning speed and achievable solution

C. Waubert de Puiseau, *Advancing Training and Inference Methods for Reinforcement Learning-Based Job Shop Scheduling*,
https://doi.org/10.1007/978-3-658-51305-4_9

quality. The proposition was to incorporate bottleneck information as inductive bias into the observation space of DRL agents, inspired by the relevance of bottlenecks in real-world systems and in the well-known shifting bottleneck heuristic (SBH). SBH-inspired bottleneck features were designed and added to several DRL-agents. However, the results showed an indifference to the addition of bottleneck information. Since the calculation of the bottleneck information generates considerable computational overhead, it is concluded that the specific method of domain knowledge integration is not recommended. Yet, the integration of domain knowledge in general remains an underexplored research area and promising avenue for future work.

The second research question focused on how to make DRL training methods more efficient and effective for the JSSP. To address this, the concept of curriculum learning (CL) was adopted from recent research in artificial intelligence (AI). In CL, the difficulty of presented tasks is adjusted over the course of training AI models. This resembles how humans are taught progressively challenging tasks in school curricula. This thesis contributed a novel metric for the difficulty of JSSP instances based on the most-tasks-remaining priority rule. This metric enables the design of curricula that vary the difficulty of training instances during training. The experimental results showed that starting training with more difficult instances consistently improved final performance, demonstrating the potential to optimize training through carefully designed curricula in DRL-based scheduling.

The third research gap was indicated by an overwhelming focus of research in the literature on elementary agent design choices for training, such as the artificial neural network architecture, rewards, observations, or DRL algorithms. Meanwhile the utilization phase of the agent received comparatively little attention. Addressing this gap, the third research question thus explored how trained DRL agents can be used most effectively during inference to generate the best possible plans within a given computation time. This was investigated in two separate studies.

The first study introduced δ-sampling, a technique that enables smooth control over the tradeoff between explorative and exploitative when generating schedules. This led to an improved coverage of the solution space given the number of solutions that may be generated due to time constrictions. An effective algorithm for determining this parameter was also proposed. The results showed that δ sampling can significantly improve the solution quality in the utilization phase without additional training. Furthermore, the method is transferable to other combinatorial problems addressed with DRL, which makes δ-sampling a valuable contribution beyond just scheduling applications.

The second study in the context of research question three proposed the use of a Decision Transformer (DT) to extract the implicit knowledge represented in a learned policy and surpass its performance. Specifically, the DT learns from interaction data of a trained base DRL agent with its environment. The DT has two advantages over the base agent. First, it explicitly incorporates the context of previous actions and their effect on the current JSSP instance into its decision. Second, it is trained to adapt its strategy to a manually adjustable goal, in this case a shorter makespan. The experiments clearly show that the DT can outperform its base DRL agents, not only but especially when longer computing times are allowed. Thus, both the DT and δ-sampling are novel and distinct ways to leverage the knowledge learned through DRL more effectively in the utilization phase.

The fourth research question addressed the implications of transitioning from dealing with academic problem settings like the JSSP to building practical scheduling systems for real-world machine scheduling problems. It was shown that embedding DRL-based scheduling systems into practice introduces additional technical and societal challenges, such as integrating more information sources and human decision-making. Additionally, several exemplary additional constraints and stochastic elements in real-world problem settings were identified that make the learning problems more difficult for DRL agents compared to the JSSP. This thesis proposed specific ideas to mitigate additional challenges and a generally applicable methodological approach to novel challenges. Special emphasis was placed on the notion of reliability, which becomes a fundamental prerequisite for scheduling systems that are used in practice. Since definitions for reliability of DRL-based scheduling systems were lacking, such definitions and associated metrics were derived for different subcategories of reliability in this domain. For each, corresponding improvement strategies were outlined, offering a basis for more structured and goal-oriented future research.

In summary, this dissertation made substantial contributions to advancing the training and inference of DRL-based JSS agents by addressing previously underexplored research directions in this interdisciplinary domain. These contributions also bring the field closer to realizing the greater vision: end-to-end DRL-based machine scheduling systems capable of optimizing economic, environmental, and societal goals in real-world production environments.

Bibliography

1. D. Silver *et al.*, "A general reinforcement learning algorithm that masters chess, shogi, and Go through self-play," *Science (New York, N.Y.)*, vol. 362, no. 6419, pp. 1140–1144, 2018, https://doi.org/10.1126/science.aar6404.
2. M. R. Garey, D. S. Johnson, and R. Sethi, "The Complexity of Flowshop and Jobshop Scheduling," *Mathematics of Operations Research*, vol. 1, no. 2, pp. 117–129, 1976, https://doi.org/10.1287/moor.1.2.117.
3. R. Haupt, "A survey of priority rule-based scheduling," *OR Spectrum*, vol. 11, no. 1, pp. 3–16, 1989, https://doi.org/10.1007/BF01721162.
4. R. Pellerin, N. Perrier, and F. Berthaut, "A survey of hybrid metaheuristics for the resource-constrained project scheduling problem," *European Journal of Operational Research*, vol. 280, no. 2, pp. 395–416, 2020, https://doi.org/10.1016/j.ejor.2019.01.063.
5. B. Naderi, R. Ruiz, and V. Roshanaei, "Mixed-Integer Programming vs. Constraint Programming for Shop Scheduling Problems: New Results and Outlook," *INFORMS Journal on Computing*, vol. 35, no. 4, pp. 817–843, 2023, https://doi.org/10.1287/ijoc.2023.1287.
6. A. Esteso, D. Peidro, J. Mula, and M. Díaz-Madroñero, "Reinforcement Learning Applied to Production Planning and Control," *International Journal of Production Research*, vol. 61, no. 16, pp. 5772–5789, 2022, https://doi.org/10.1080/00207543.2022.2104180.
7. B. Waschneck *et al.*, "Optimization of global production scheduling with deep reinforcement learning," *Procedia CIRP*, vol. 72, pp. 1264–1269, 2018, https://doi.org/10.1016/j.procir.2018.03.212.
8. X. Wang, Y. Chen, and W. Zhu, "A Survey on Curriculum Learning," *IEEE Transactions on Pattern Analysis and Machine Intelligence*, vol. 44, no. 9, pp. 4555–4576, 2022, https://doi.org/10.1109/TPAMI.2021.3069908.

C. Waubert de Puiseau, *Advancing Training and Inference Methods for Reinforcement Learning-Based Job Shop Scheduling*,
https://doi.org/10.1007/978-3-658-51305-4

9. J. Adams, E. Balas, and D. Zawack, "The Shifting Bottleneck Procedure for Job Shop Scheduling," *Management Science*, vol. 34, no. 3, pp. 391–401, 1988, https://doi.org/10.1287/mnsc.34.3.391.
10. H. Lödding, *Handbook of Manufacturing Control,* 1st ed. Berlin, Heidelberg: Springer Berlin Heidelberg, 2013.
11. M. Pinedo, *Scheduling: Theory, algorithms, and systems,* 5th ed.: Springer International Publishing, 2016.
12. H. Luczak and W. Eversheim, *Produktionsplanung und -steuerung: Grundlagen, Gestaltung und Konzepte,* 2nd ed. Berlin, Heidelberg: Springer Berlin Heidelberg, 1999.
13. R. L. Graham, E. L. Lawler, J. K. Lenstra, and A. Kan, "Optimization and Approximation in Deterministic Sequencing and Scheduling: a Survey," in *Annals of Discrete Mathematics, Discrete Optimization II, Proceedings of the Advanced Research Institute on Discrete Optimization and Systems Applications of the Systems Science Panel of NATO and of the Discrete Optimization Symposium co-sponsored by IBM Canada and SIAM Banff, Aha. and Vancouver*: Elsevier, 1979, pp. 287–326.
14. A. Rinciog and A. Meyer, "Towards Standardizing Reinforcement Learning Approaches for Stochastic Production Scheduling," *Procedia CIRP*, vol. 107, pp. 1112–1119, 2022.
15. V. Samsonov, M. Behery, and G. Lakemeyer, "Reinforcement Learning for Short-Term Production Scheduling with Sequence-Dependent Setup Waste," *ERCIM NEWS*, no. 122, pp. 38–39, 2020.
16. J. Popper, W. Motsch, A. David, T. Petzsche, and M. Ruskowski, "Utilizing multi-agent deep reinforcement learning for flexible job shop scheduling under sustainable viewpoints," in *2021 International Conference on Electrical, Computer, Communications and Mechatronics Engineering (ICECCME)*, Mauritius, Mauritius, 07.10.2021–08.10.2021, 2021, pp. 1–6, https://doi.org/10.1109/ICECCME52200.2021.9590925.
17. W.-Y. Ku and J. C. Beck, "Mixed Integer Programming models for job shop scheduling: A computational analysis," *Computers & Operations Research*, vol. 73, pp. 165–173, 2016, https://doi.org/10.1016/j.cor.2016.04.006.
18. J. Błażewicz, E. Pesch, and M. Sterna, "The disjunctive graph machine representation of the job shop scheduling problem," *European Journal of Operational Research*, vol. 127, no. 2, pp. 317–331, 2000, https://doi.org/10.1016/S0377-2217(99)00486-5.
19. I. G. Smit *et al.*, "Graph neural networks for job shop scheduling problems: A survey," *Computers & Operations Research*, vol. 176, p. 106914, 2025, https://doi.org/10.1016/j.cor.2024.106914.
20. B. Korte and J. Vygen, *Combinatorial Optimization.* Berlin, Heidelberg: Springer Berlin Heidelberg, 2018.
21. J. Ostrowski, M. F. Anjos, and A. Vanelli, "Symmetry in scheduling problems," Departement de mathématiques et de génie industriel, École Polytechnique de Montréal, Montreal, 2010.
22. E. Taillard, "Benchmarks for basic scheduling problems," *European Journal of Operational Research*, vol. 64, no. 2, pp. 278–285, 1993, https://doi.org/10.1016/0377-2217(93)90182-m.
23. G. Da Col and E. C. Teppan, "Industrial Size Job Shop Scheduling Tackled by Present Day CP Solvers," in *Lecture Notes in Computer Science, Principles and Practice of*

Constraint Programming: 25th international, T. Schiex and S. de Givry, Eds.: Springer Nature, 2019, pp. 144–160.
24. H. Fischer and G. L. Thompson, "Probabilistic learning combinations of local job-shop scheduling rules," *Prentice Hall*, pp. 225–251, 1963.
25. S. Lawrence, "Resource Constrained Project Scheduling. An Experimental Investigation of Heuristic Scheduling Techniques (Supplement)," *Graduate School of Industrial Administration, Carnegie-Mellon University*, 1984.
26. D. Applegate and W. Cook, "A Computational Study of the Job-Shop Scheduling Problem," *ORSA Journal on Computing*, vol. 3, no. 2, pp. 149–156, 1991, https://doi.org/10.1287/ijoc.3.2.149.
27. R. H. Storer, S. D. Wu, and R. Vaccari, "New Search Spaces for Sequencing Problems with Application to Job Shop Scheduling," *Management Science*, vol. 38, no. 10, pp. 1495–1509, 1992, https://doi.org/10.1287/mnsc.38.10.1495.
28. T. Yamada and R. Nakano, "A Genetic Algorithm Applicable to Large-Scale Job-Shop Problems," in *Parallel Problem Solving from Nature 2, PPSN-II*, Brussels, Belgium, September 28–30, 1992, pp. 281–290.
29. E. Demirkol, S. Mehta, and R. Uzsoy, "A Computational Study of Shifting Bottleneck Procedures for Shop Scheduling Problems," *Journal of Heuristics*, vol. 3, no. 2, pp. 111–137, 1997, https://doi.org/10.1023/A:1009627429878.
30. S. R. Hong, J. Hullman, and E. Bertini, "Human Factors in Model Interpretability: Industry Practices, Challenges, and Needs," *Proceedings of the ACM on Human-Computer Interaction*, vol. 4, no. 1, pp. 1–26, 2020, https://doi.org/10.1145/3392878.
31. R. Martí, P. M. Pardalos, and M. G. C. Resende, *Handbook of Heuristics*. Cham: Springer International Publishing, 2018.
32. M. Kemmerling, "Job shop scheduling with neural Monte Carlo Tree Search," RWTH Aachen University, 2024.
33. Gurobi, *Gurobi.* [Online]. Available: https://www.gurobi.com/ (accessed: Oct. 15 2024).
34. L. Perron and F. Didier, *CP-SAT.* [Online]. Available: https://developers.google.com/optimization/cp/cp_solver/
35. M. Gendreau and J.-Y. Potvin, *Handbook of Metaheuristics,* 1st ed. Cham: Springer International Publishing, 2019.
36. F. Glover, "Future paths for integer programming and links to artificial intelligence," *Computers & Operations Research*, vol. 13, no. 5, pp. 533–549, 1986, https://doi.org/10.1016/0305-0548(86)90048-1.
37. S. Kirkpatrick, C. D. Gelatt, and M. P. Vecchi, "Optimization by simulated annealing," *Science (New York, N.Y.)*, vol. 220, no. 4598, pp. 671–680, 1983, https://doi.org/10.1126/science.220.4598.671.
38. D. E. Goldberg, *Genetic algorithms in search, optimization, and machine learning,* 30th ed. Boston: Addison-Wesley, 2012.
39. J. Kennedy and R. Eberhart, "Particle swarm optimization," in *Proceedings / 1995 IEEE International Conference on Neural Networks*, Perth, WA, Australia, 27 Nov.-1 Dec. 1995, Nov. 1995, pp. 1942–1948, https://doi.org/10.1109/ICNN.1995.488968.
40. T. van Ekeris, R. Meyes, and T. Meisen, "Discovering Heuristics And Metaheuristics For Job Shop Scheduling From Scratch Via Deep Reinforcement Learning," in *Proceedings of the International Conference on Production Systems and Logistics:*

CPSL 2021, Digital Event, August 10–11, 2021, pp. 709–718, https://doi.org/10.15488/11231.

41. P. Tassel, M. Gebser, and K. Schekotihin, "An End-to-End Reinforcement Learning Approach for Job-Shop Scheduling Problems Based on Constraint Programming," in *Proceedings of the Thirty-Third International Conference on Automated Planning and Scheduling*, Prague, Czech Republic, July 8–13, 2023, pp. 614–622.
42. J. K. Falkner, D. Thyssens, A. Bdeir, and L. Schmidt-Thieme, "Learning to Control Local Search for Combinatorial Optimization," in *Machine Learning and Knowledge Discovery in Databases—European Conference*, Genoble, France, September 19–23, 2022, pp. 361–376, https://doi.org/10.1007/978-3-031-26419-1_22.
43. J. Kallestad, R. Hasibi, A. Hemmati, and K. Sörensen, "A general deep reinforcement learning hyperheuristic framework for solving combinatorial optimization problems," *European Journal of Operational Research*, vol. 309, no. 1, pp. 446–468, 2023, https://doi.org/10.1016/j.ejor.2023.01.017.
44. R. S. Sutton and A. Barto, *Reinforcement learning: An introduction,* 2nd ed. Cambridge, Massachusetts, London, England: The MIT Press, 2018.
45. R. Bellman, "The theory of dynamic programming," *Bulletin of the American Mathematical Society*, vol. 60, no. 6, pp. 503–515, 1954.
46. I. Goodfellow, Y. Bengio, and A. Courville, *Deep Learning*: MIT Press, 2016.
47. J. Achiam, *Spinning Up in Deep Reinforcement Learning.* [Online]. Available: https://spinningup.openai.com/en/latest/index.html (accessed: May 10 2025).
48. H. van Hasselt, A. Guez, and D. Silver, "Deep Reinforcement Learning with Double Q-Learning," in *Proceedings of the Thirtieth AAAI Conference on Artificial Intelligence*, Phoenix, Arizona, USA, February 12–17, 2016, pp. 2094–2100, https://doi.org/10.1609/aaai.v30i1.10295.
49. R. J. Williams, "Simple statistical gradient-following algorithms for connectionist reinforcement learning," *Machine Learning*, vol. 8, 3–4, pp. 229–256, 1992, https://doi.org/10.1007/BF00992696.
50. V. Mnih *et al.,* "Asynchronous Methods for Deep Reinforcement Learning," in *Proceedings of the 33nd International Conference on Machine Learning*, New York City, NY, USA, June 19–24, 2016, pp. 1928–1937.
51. J. Schulman, S. Levine, P. Moritz, M. I. Jordan, and P. Abbeel, "Trust Region Policy Optimization," in *Proceedings of the 32nd International Conference on Machine Learning*, Lille France, July 6–11, 2015, pp. 1889–1897.
52. J. Schulman, F. Wolski, P. Dhariwal, A. Radford, and O. Klimov, "Proximal Policy Optimization Algorithms," Jul. 2017. [Online]. Available: http://arxiv.org/pdf/1707.06347v2
53. P.-L. Bacon, J. Harb, and D. Precup, "The Option-Critic Architecture," in *Proceedings of the Thirty-First AAAI Conference on Artificial Intelligence*, San Francisco, California, USA, February 4–9, 2017, pp. 1726–1734.
54. T. P. Lillicrap *et al.,* "Continuous control with deep reinforcement learning," in *4th International Conference on Learning Representations*, San Juan, Puerto Rico, May 2–4, 2016.
55. T. Salimans, J. Ho, X. Chen, S. Sidor, and I. Sutskever, "Evolution Strategies as a Scalable Alternative to Reinforcement Learning," Mar. 2017. [Online]. Available: http://arxiv.org/pdf/1703.03864v2

56. A. Corsini, A. Porrello, S. Calderara, and M. Dell'Amico, "Self-Labeling the Job Shop Scheduling Problem," in *Advances in Neural Information Processing Systems 38: Annual on Neural Information Processing Systems 2024*, Vancouver, BC, Canada, December 10–15, 2024, 105528–105551.
57. J. Pirnay and D. G. Grimm, "Self-Improvement for Neural Combinatorial Optimization: Sample without Replacement, but Improvement," *Transaction on Machine Learning Research*, pp. 2835–8856, 2024. [Online]. Available: https://openreview.net/forum?id=agT8ojoH0X
58. V. Samsonov, K. B. Hicham, and T. Meisen, "Reinforcement Learning in Manufacturing Control: Baselines, Challenges and Ways Forward," *Engineering Applications of Artificial Intelligence*, vol. 112, p. 104868, 2022, https://doi.org/10.1016/j.engappai.2022.104868.
59. A. Kuhnle, J.-P. Kaiser, F. Theiß, N. Stricker, and G. Lanza, "Designing an adaptive production control system using reinforcement learning," *Journal of Intelligent Manufacturing*, no. 44, pp. 855–876, 2021, https://doi.org/10.1007/s10845-020-01612-y.
60. J. Bjorck, C. P. Gomes, and K. Q. Weinberger, "Towards Deeper Deep Reinforcement Learning with Spectral Normalization," in *Advances in Neural Information Processing Systems 34: Annual Conference on Neural Information Processing Systems 2021*, Virtual, December 6–14, 2021, pp. 8242–8255.
61. A. Raffin, A. Hill, A. Gleave, A. Kanervisto, M. Ernestus, and N. Dormann, "Stable-Baselines3: Reliable Reinforcement Learning Implementations," *Journal of Machine Learning Research*, vol. 22, no. 268, pp. 1–8, 2021. [Online]. Available: http://jmlr.org/papers/v22/20-1364.html
62. C. Waubert de Puiseau, R. Meyes, and T. Meisen, "On reliability of reinforcement learning based production scheduling systems: a comparative survey," *Journal of Intelligent Manufacturing*, vol. 33, no. 4, pp. 911–927, 2022, https://doi.org/10.1007/s10845-022-01915-2.
63. F. Fruggiero, S. Riemma, Y. Ouazene, R. Macchiaroli, and V. Guglielmi, "Incorporating the Human Factor within Manufacturing Dynamics," *IFAC-PapersOnLine*, vol. 49, no. 12, pp. 1691–1696, 2016, https://doi.org/10.1016/j.ifacol.2016.07.825.
64. X. Cai, Y. Lu, and J. Wang, "The impact of temperature on manufacturing worker productivity: Evidence from personnel data," *Journal of Comparative Economics*, vol. 46, no. 4, pp. 889–905, 2018, https://doi.org/10.1016/j.jce.2018.06.003.
65. O. Lohse, N. Pütz, and K. Hörmann, "Implementing an Online Scheduling Approach for Production with Multi Agent Proximal Policy Optimization (MAPPO)," in *Advances in Production Management Systems. Artificial Intelligence for Sustainable and Resilient Production Systems*, Nantes, France, September 5–9, 2021, pp. 586–595.
66. G. Bonetta, D. Zago, R. Cancelliere, and A. Grosso, "Job Shop Scheduling via Deep Reinforcement Learning: A Sequence to Sequence Approach," in *Lecture Notes in Computer Science*, vol. 14286, *Learning and intelligent optimization: 17th international conference, LION 17, Nice, France, June 4–8, 2023 : revised selected papers*, M. Sellmann and K. Tierney, Eds., Cham: Springer, 2023, pp. 475–490.
67. C. Zhang, W. Song, Z. Cao, J. Zhang, P. S. Tan, and X. Chi, "Learning to Dispatch for Job Shop Scheduling via Deep Reinforcement Learning," in *Advances in Neural Information Processing Systems 33: Annual Conference on Neural Information Processing*

Systems 2020, Virtual, December 6–12, 2020, pp. 1621–1632.

68. K.-H. Ho, R.-Y. Jheng, J.-H. Wu, F. Chiang, and Y.-C. Chen, “Residual Scheduling: A New Reinforcement Learning Approach to Solving Job Shop Scheduling Problem,” *IEEE Access*, vol. 12, 14703–14718, 2024, https://doi.org/10.1109/ACCESS.2024.3357969.
69. J. Park, J. Chun, S. H. Kim, Y. Kim, and J. Park, “Learning to schedule job-shop problems: representation and policy learning using graph neural network and reinforcement learning,” *International Journal of Production Research*, vol. 59, no. 11, pp. 3360–3377, 2021, https://doi.org/10.1080/00207543.2020.1870013.
70. P. Tassel, M. Gebser, and K. Schekotihin, “A Reinforcement Learning Environment For Job-Shop Scheduling,” arXiv:2104.03760, Apr. 2021. [Online]. Available: https://arxiv.org/pdf/2104.03760
71. K.-H. Ho *et al.*, “Deep Reinforcement Learning Based on Graph Neural Networks for Job-shop Scheduling,” *2023 International Conference on Consumer Electronics—Taiwan, ICCE-Taiwan 2023—Proceedings*, 2023, https://doi.org/10.1109/ICCE-Taiwan58799.2023.10226873.
72. W. Song, X. Chen, Q. Li, and Z. Cao, “Flexible Job Shop Scheduling via Graph Neural Network and Deep Reinforcement Learning,” *IEEE Transactions on Industrial Informatics*, pp. 1–11, 2022, https://doi.org/10.1109/TII.2022.3189725.
73. C. Waubert de Puiseau, H. Tercan, and T. Meisen, “Curriculum Learning in Job Shop Scheduling using Reinforcement Learning,” in *Proceedings of the Conference on Production Systems and Logistics: CPSL 2023*, Querétaro, Mexico, February 28—March 2, 2023, pp. 34–43, https://doi.org/10.15488/13422.
74. Z. Iklassov, D. Medvedev, R. Solozabal, and M. Takac, “On the Study of Curriculum Learning for Inferring Dispatching Policies on the Job Shop Scheduling,” in *Proceedings of the Thirty-Second International Joint Conference on Artificial Intelligence*, Macao, SAR, China, August 19–25, 2023, pp. 5350–5358.
75. R. Chen, W. Li, and H. Yang, “A Deep Reinforcement Learning Framework Based on an Attention Mechanism and Disjunctive Graph Embedding for the Job-Shop Scheduling Problem,” *IEEE Transactions on Industrial Informatics*, vol. 19, no. 2, pp. 1322–1331, 2022, https://doi.org/10.1109/TII.2022.3167380.
76. M. Felder *et al.*, “Energy-Flexible Job-Shop Scheduling Using Deep Reinforcement Learning,” *Proceedings of the Conference on Production Systems and Logistics*, 2023, https://doi.org/10.15488/13454.
77. M. S. Martins *et al.*, “Reinforcement learning for dual-resource constrained scheduling,” *IFAC-PapersOnLine*, vol. 53, 2020, https://doi.org/10.1016/j.ifacol.2020.12.2866.
78. J. Pirnay, Q. Göttl, J. Burger, and D. G. Grimm, “Policy-Based Self-Competition for Planning Problems,” in *The Eleventh International Conference on Learning Representations*, Kigali, Rwanda, May 1–5, 2023, pp. 1–24. [Online]. Available: https://openreview.net/forum?id=SmufNDN90G
79. L. Zhao, W. Shen, C. Zhang, and K. Peng, “An End-to-End Deep Reinforcement Learning Approach for Job Shop Scheduling,” in *Proceedings of the 2022 IEEE 25th International Conference on Computer Supported Cooperative Work in Design (CSCWD): May 4–6, 2022, Hangzhou, China*, Hangzhou, China, 5/4/2022–5/6/2022, 2022, pp. 841–846, https://doi.org/10.1109/CSCWD54268.2022.9776116.

80. S. Qu, J. Wang, and J. Jasperneite, "Dynamic scheduling in large-scale stochastic processing networks for demand-driven manufacturing using distributed reinforcement learning," in *2018 IEEE 23rd International Conference on Emerging Technologies and Factory Automation (ETFA)*, Turin, Italy, 04.09.2018–07.09.2018, 2018, pp. 433–440, https://doi.org/10.1109/ETFA.2018.8502508.
81. Z. Liao, J. Chen, and Z. Zhang, "Solving Job-Shop Scheduling Problem via Deep Reinforcement Learning with Attention Model," *Lecture Notes in Computer Science (including subseries Lecture Notes in Artificial Intelligence and Lecture Notes in Bioinformatics)*, 13926 LNAI, pp. 201–212, 2023, https://doi.org/10.1007/978-3-031-36822-6_18.
82. D. Vivekanandan, S. Wirth, P. Karlbauer, and N. Klarmann, "A Reinforcement Learning Approach for Scheduling Problems With Improved Generalization Through Order Swapping," *Machine Learning and Knowledge Extraction*, vol. 5, no. 2, pp. 418–430, 2023.
83. E. K. Elsayed, A. K. Elsayed, and K. A. Eldahshan, "Deep Reinforcement Learning-Based Job Shop Scheduling of Smart Manufacturing," *Computers, Materials and Continua*, vol. 73, no. 3, pp. 5103–5120, 2022, https://doi.org/10.32604/cmc.2022.030803.
84. B. Cunha, A. Madureira, B. Fonseca, and J. Matos, "Intelligent scheduling with reinforcement learning," *Applied Sciences (Switzerland)*, vol. 11, no. 8, 2021, https://doi.org/10.3390/app11083710.
85. C.-L. Liu, C.-C. Chang, and C.-J. Tseng, "Actor-Critic Deep Reinforcement Learning for Solving Job Shop Scheduling Problems," *IEEE Access*, vol. 8, pp. 71752–71762, 2020, https://doi.org/10.1109/ACCESS.2020.2987820.
86. S.-W. Jeon, D. Lee, S.-C. Oh, K.-T. Park, S.-D. Noh, and J. Arinez, "Design and Implementation of Simulation-Based Scheduling System with Reinforcement Learning for Re-Entrant Production Lines," *Machines*, vol. 10, no. 12, 2022, https://doi.org/10.3390/machines10121169.
87. S. Abderrazzak, A. Hamid, and S. Omar, "Stochastic Dynamic Programming for Earliness-Tardiness Single Machine Scheduling with Maintenance Considerations," *IFIP Advances in Information and Communication Technology*, no. 634, pp. 269–276, 2021.
88. M. S. A. Hameed and A. Schwung, "Graph neural networks-based Scheduler for Production planning problems using Reinforcement Learning," *Journal of Manufacturing Systems*, vol. 69, pp. 91–102, 2023.
89. J. Park, S. Bakhtiyar, and J. Park, "ScheduleNet: Learn to solve multi-agent scheduling problems with reinforcement learning," Jun. 2021. [Online]. Available: http://arxiv.org/pdf/2106.03051v1
90. B. Waschneck *et al.*, "Deep reinforcement learning for semiconductor production scheduling," *2018 29th Annual SEMI Advanced Semiconductor Manufacturing Conference, ASMC 2018*, 2018, https://doi.org/10.1109/ASMC.2018.8373191.
91. L. B. Wang *et al.*, "Dynamic job-shop scheduling in smart manufacturing using deep reinforcement learning," *Computer Networks*, vol. 190, 107969, 2021, https://doi.org/10.1016/j.comnet.2021.107969.
92. Z. Liao, Q. Li, Y. Dai, and Z. Zhang, "Learning to Schedule Job-Shop Problems via Hierarchical Reinforcement Learning," *Proceedings—2022 International conference*

on Systems, Man, and Cybernetics (SMC), pp. 3222–3227, 2022, https://doi.org/10.1109/SMC53654.2022.9945585.

93. L. Li and L. Lin, "Scheduling Distributed Flexible Assembly Lines using Safe Reinforcement Learning with Soft Shielding," in *2024 IEEE 7th Advanced Information Technology, Electronic and Automation Control Conference (IAEAC)*, Chongqing, China, 15.03.2024–17.03.2024, 2024, pp. 646–652, https://doi.org/10.1109/IAEAC59436.2024.10503804.
94. J. C. Serrano-Ruiz, J. Mula, and R. Poler, "Job shop smart manufacturing scheduling by deep reinforcement learning," *Journal of Industrial Information Integration*, vol. 38, p. 100582, 2024, https://doi.org/10.1016/j.jii.2024.100582.
95. J.-P. Huang, L. Gao, X.-Y. Li, and C.-J. Zhang, "A novel priority dispatch rule generation method based on graph neural network and reinforcement learning for distributed job-shop scheduling," *Journal of Manufacturing Systems*, vol. 69, pp. 119–134, 2023, https://doi.org/10.1016/j.jmsy.2023.06.007.
96. C. Su *et al.*, "Evolution strategies-based optimized graph reinforcement learning for solving dynamic job shop scheduling problem," *Applied Soft Computing*, vol. 145, p. 110596, 2023, https://doi.org/10.1016/j.asoc.2023.110596.
97. V. Samsonov *et al.*, "Manufacturing Control in Job Shop Environments with Reinforcement Learning," in *Proceedings of the 13th International Conference on Agents and Artificial Intelligence*, online, February 4–6, 2021, pp. 589–597, https://doi.org/10.5220/0010202405890597.
98. M. Monaci, V. Agasucci, and G. Grani, "An actor-critic algorithm with policy gradients to solve the job shop scheduling problem using deep double recurrent agents," *European Journal of Operational Research*, vol. 312, no. 3, pp. 910–926, 2024, https://doi.org/10.1016/j.ejor.2023.07.037.
99. J. Ahn, S. Yun, J.-W. Kwon, and W.-T. Kim, "Literacy Deep Reinforcement Learning-Based Federated Digital Twin Scheduling for the Software-Defined Factory," *Electronics (Switzerland)*, vol. 13, no. 22, p. 4452, 2024, https://doi.org/10.3390/electronics13224452.
100. K. Alexopoulos, P. Mavrothalassitis, E. Bakopoulos, N. Nikolakis, and D. Mourtzis, "Deep Reinforcement Learning for Selection of Dispatch Rules for Scheduling of Production Systems," *Applied Sciences*, vol. 15, no. 1, p. 232, 2025, https://doi.org/10.3390/app15010232.
101. R. Maharjan, P.-A. Andersen, and L. Jiao, "Exploring Efficient Job Shop Scheduling Using Deep Reinforcement Learning," *International Conference on Innovative Techniques and Applications of Artificial Intelligence*, pp. 251–257, 2025, https://doi.org/10.1007/978-3-031-77918-3_20.
102. C. Schmidl, T. D. Simão, and N. Jansen, "A Supervised Learning Approach to Robust Reinforcement Learning for Job Shop Scheduling," in *Proceedings of the 16th International Conference on Agents and Artificial Intelligence*, Rome, Italy, February 24–26, 2024, 1324–1335. [Online]. Available: https://www.scopus.com/inward/record.uri?eid=2-s2.0-85190804416&doi=10.5220%2f0012473600003636&partnerID=40&md5=ee54d2d797f3a2ab4922810777a21eb4#, https://doi.org/10.5220/0012473600003636.
103. C.-C. Lin, D.-J. Deng, Y.-L. Chih, and H.-T. Chiu, "Smart Manufacturing Scheduling With Edge Computing Using Multiclass Deep Q Network," *IEEE Transactions on*

Industrial Informatics, vol. 15, no. 7, pp. 4276–4284, 2019, https://doi.org/10.1109/TII.2019.2908210.

104. Y. Zeng, Z. Liao, Y. Dai, R. Wang, X. Li, and B. Yuan, "Hybrid intelligence for dynamic job-shop scheduling with deep reinforcement learning and attention mechanism," arXiv:2201.00548, Jan. 2022. [Online]. Available: https://arxiv.org/pdf/
105. T. Zhou, D. Tang, H. Zhu, and Z. Zhang, "Multi-agent reinforcement learning for online scheduling in smart factories," *Robotics and Computer-Integrated Manufacturing*, vol. 72, p. 102202, 2021, https://doi.org/10.1016/j.rcim.2021.102202.
106. T. Zhou *et al.*, "Reinforcement learning for online optimization of job-shop scheduling in a smart manufacturing factory," *Advances in Mechanical Engineering*, vol. 14, no. 3, 2022, https://doi.org/10.1177/16878132221086120.
107. T. Altenmüller, T. Stüker, B. Waschneck, A. Kuhnle, and G. Lanza, "Reinforcement learning for an intelligent and autonomous production control of complex job-shops under time constraints," *Production Engineering: Research and Development*, vol. 14, no. 3, pp. 319–328, 2020, https://doi.org/10.1007/s11740-020-00967-8.
108. Y. Turgut and C. E. Bozdag, "Deep Q-network Model For Dynamic Job Shop Scheduling Pproblem Based On Discrete Event Simulation," in *2020 Winter Simulation Conference (WSC)*, 2020, 1551–1559.
109. M. Zhang, Y. Lu, Y. X. Hu, N. Amaitik, and Y. C. Xu, "Dynamic Scheduling Method for Job-Shop Manufacturing Systems by Deep Reinforcement Learning with Proximal Policy Optimization," *Sustainability*, vol. 14, no. 9, 2022, https://doi.org/10.3390/su14095177.
110. B. Luo, S. Wang, B. Yang, and L. Yi, "An improved deep reinforcement learning approach for the dynamic job shop scheduling problem with random job arrivals," *Journal of Physics: Conference Series*, vol. 1848, no. 1, p. 12029, 2021, https://doi.org/10.1088/1742-6596/1848/1/012029.
111. Z. Wang and W. Liao, "Smart scheduling of dynamic job shop based on discrete event simulation and deep reinforcement learning," *Journal of Intelligent Manufacturing*, 2023, https://doi.org/10.1007/s10845-023-02161-w.
112. Z. Yang, L. Bi, and X. Jiao, "Combining Reinforcement Learning Algorithms with Graph Neural Networks to Solve Dynamic Job Shop Scheduling Problems," *Processes*, vol. 11, no. 5, p. 1571, 2023, https://doi.org/10.3390/pr11051571.
113. B.-A. Han and J.-J. Yang, "Research on Adaptive Job Shop Scheduling Problems Based on Dueling Double DQN," *IEEE Access*, vol. 8, pp. 186474–186495, 2020, https://doi.org/10.1109/ACCESS.2020.3029868.
114. C. -L. Liu, C. -J. Tseng, and P. -H. Weng, "Dynamic Job-Shop Scheduling via Graph Attention Networks and Deep Reinforcement Learning," *IEEE Transactions on Industrial Informatics*, vol. 20, no. 6, pp. 8662–8672, 2024, https://doi.org/10.1109/TII.2024.3371489.
115. C. -L. Liu and T. -H. Huang, "Dynamic Job-Shop Scheduling Problems Using Graph Neural Network and Deep Reinforcement Learning," *IEEE Transactions on Systems, Man, and Cybernetics: Systems*, vol. 53, no. 11, pp. 6836–6848, 2023, https://doi.org/10.1109/TSMC.2023.3287655.
116. Y. Lei, Q. Deng, M. Liao, and S. Gao, "Deep reinforcement learning for dynamic distributed job shop scheduling problem with transfers," *Expert Systems with Applications*, vol. 251, p. 123970, 2024, https://doi.org/10.1016/j.eswa.2024.123970.

117. Z. Liu, H. Mao, G. Sa, H. Liu, and J. Tan, "Dynamic job-shop scheduling using graph reinforcement learning with auxiliary strategy," *Journal of Manufacturing Systems*, vol. 73, pp. 1–18, 2024, https://doi.org/10.1016/j.jmsy.2024.01.002.
118. X. Wu, X. Yan, D. Guan, and M. Wei, "A deep reinforcement learning model for dynamic job-shop scheduling problem with uncertain processing time," *Engineering Applications of Artificial Intelligence*, vol. 131, p. 107790, 2024, https://doi.org/10.1016/j.engappai.2023.107790.
119. Y. J. Zhao, Y. H. Wang, Y. Y. Tan, J. Zhang, and H. X. Yu, "Dynamic Jobshop Scheduling Algorithm Based on Deep Q Network," *IEEE Access*, vol. 9, pp. 122995–123011, 2021, https://doi.org/10.1109/ACCESS.2021.3110242.
120. R. Liu, R. Piplani, and C. Toro, "A deep multi-agent reinforcement learning approach to solve dynamic job shop scheduling problem," *Computers & Operations Research*, vol. 159, p. 106294, 2023, https://doi.org/10.1016/j.cor.2023.106294.
121. L. Song, Y. Li, and J. Xu, "Dynamic Job-Shop Scheduling Based on Transformer and Deep Reinforcement Learning," *Processes*, vol. 11, no. 12, p. 3434, 2023, https://doi.org/10.3390/pr11123434.
122. C. Chen, C. Yin, and R. Wang, "Dynamic job shop scheduling method based on ensemble reinforcement learning," in *2024 6th International Conference on Internet of Things, Automation and Artificial Intelligence (IoTAAI)*, Guangzhou, China, July 26–28, 2024, https://doi.org/10.1109/IoTAAI62601.2024.10692885.
123. S. H. Oh, Y. in Cho, and J. H. Woo, "Distributional reinforcement learning with the independent learners for flexible job shop scheduling problem with high variability," *Journal of Computational Design and Engineering*, vol. 9, no. 4, pp. 1157–1174, 2022, https://doi.org/10.1093/jcde/qwac044.
124. S. Pol, S. Baer, D. Turner, V. Samsonov, and T. Meisen, "Global Reward Design for Cooperative Agents to Achieve Flexible Production Control under Real-time Constraints," *ICEIS 2021: Proceedings of the 23rd International Conference on Enterprise Information Systems*, vol. 1, pp. 515–526, 2021, https://doi.org/10.5220/0010455805150526.
125. J. Popper and M. Ruskowski, "Using Multi-Agent Deep Reinforcement Learning For Flexible Job Shop Scheduling Problems," *Procedia CIRP*, vol. 112, pp. 63–67, 2022, https://doi.org/10.1016/j.procir.2022.09.039.
126. P. Burggräf, J. Wagner, T. Saßmannshausen, D. Ohrndorf, and K. Subramani, "Multi-agent-based deep reinforcement learning for dynamic flexible job shop scheduling," *Procedia CIRP*, vol. 112, pp. 57–62, 2022, https://doi.org/10.1016/j.procir.2022.09.024.
127. D. Johnson, G. Chen, and Y. Lu, "Multi-Agent Reinforcement Learning for Real-Time Dynamic Production Scheduling in a Robot Assembly Cell," *IEEE Robotics and Automation Letters*, vol. 7, no. 3, pp. 7684–7691, 2022, https://doi.org/10.1109/LRA.2022.3184795.
128. S. Lang, N. Lanzerath, F. Behrendt, T. Reggelin, and M. Müller, "Integration Of Deep Reinforcement Learning And Discrete-event Simulation For Real-time Scheduling Of A Flexible Job Shop Production," in *2020 Winter Simulation Conference (WSC)*, 2020, pp. 3057–3068.
129. S. H. Oh, Y. in Cho, and J. H. Woo, "Applying Multi-agent Reinforcement Learning and Graph Neural Networks to Flexible Job Shop Scheduling Problem," *IFIP Advances*

in Information and Communication Technology, 691 AICT, pp. 506–519, 2023, https://doi.org/10.1007/978-3-031-43670-3_35.
130. I. B. Park and J. Park, "Scalable Scheduling of Semiconductor Packaging Facilities Using Deep Reinforcement Learning," *IEEE Transactions on Cybernetics*, vol. 53, no. 6, pp. 3518–3531, 2023, https://doi.org/10.1109/TCYB.2021.3128075.
131. H. Zhu, S. Tao, Y. Gui, and Q. Cai, "Research on an Adaptive Real-Time Scheduling Method of Dynamic Job-Shop Based on Reinforcement Learning," *Machines*, vol. 10, no. 11, 2022, https://doi.org/10.3390/machines10111078.
132. X. Zhu, J. Xu, J. Ge, Y. Wang, and Z. Xie, "Multi-Task Multi-Agent Reinforcement Learning for Real-Time Scheduling of a Dual-Resource Flexible Job Shop with Robots," *Processes*, vol. 11, no. 1, 2023, https://doi.org/10.3390/pr11010267.
133. K. Lei *et al.*, "A multi-action deep reinforcement learning framework for flexible Job-shop scheduling problem," *Expert Systems with Applications*, vol. 205, 2022, https://doi.org/10.1016/j.eswa.2022.117796.
134. I.-B. Park, J. Huh, J. Kim, and J. Park, "A Reinforcement Learning Approach to Robust Scheduling of Semiconductor Manufacturing Facilities," *IEEE Transactions on Automation Science and Engineering*, vol. 17, no. 3, pp. 1–12, 2020, https://doi.org/10.1109/TASE.2019.2956762.
135. Z. Chen, L. Zhang, X. Wang, and P. Gu, "Optimal Design of Flexible Job Shop Scheduling Under Resource Preemption Based on Deep Reinforcement Learning," *Complex System Modeling and Simulation*, vol. 2, no. 2, pp. 174–185, 2022, https://doi.org/10.23919/CSMS.2022.0007.
136. Y. Du, J. Li, C. Li, and P. Duan, "A Reinforcement Learning Approach for Flexible Job Shop Scheduling Problem With Crane Transportation and Setup Times," *IEEE Transactions on Neural Networks and Learning Systems*, pp. 1–15, 2022, https://doi.org/10.1109/TNNLS.2022.3208942.
137. S. Wang, J. Li, H. Tang, and J. Wang, "CEA-FJSP: Carbon emission-aware flexible job-shop scheduling based on deep reinforcement learning," *Frontiers in Environmental Science*, vol. 10, 2022, https://doi.org/10.3389/fenvs.2022.1059451.
138. Y. H. Lee and S. Lee, "Deep reinforcement learning based scheduling within production plan in semiconductor fabrication," *Expert Systems with Applications*, vol. 191, 2022, https://doi.org/10.1016/j.eswa.2021.116222.
139. Z. Zeng, X. Li, and C. Bai, "A Deep Reinforcement Learning Approach to Flexible Job Shop Scheduling," in *2022 IEEE International Conference on Systems, Man, and Cybernetics (SMC)*, Prague, Czech Republic, Oktober 9–12, 2022, pp. 884–890, https://doi.org/10.1109/SMC53654.2022.9945107.
140. R. Magalhães, M. Martins, S. Vieira, F. Santos, and J. Sousa, "Encoder-Decoder Neural Network Architecture for solving Job Shop Scheduling Problems using Reinforcement Learning," in *2021 IEEE Symposium Series on Computational Intelligence (SSCI)*, 2021, pp. 1–8, https://doi.org/10.1109/SSCI50451.2021.9659849.
141. D. Huang, H. Zhao, L. Zhang, and K. Chen, "Learning to Dispatch for Flexible Job Shop Scheduling Based on Deep Reinforcement Learning via Graph Gated Channel Transformation," *IEEE Access*, vol. 12, pp. 50935–50948, 2024, https://doi.org/10.1109/ACCESS.2024.3384923.

142. R. Wang, G. Wang, J. Sun, F. Deng, and J. Chen, "Flexible Job Shop Scheduling via Dual Attention Network-Based Reinforcement Learning," *IEEE Transactions on Neural Networks and Learning Systems*, vol. 35, no. 3, pp. 3091–3102, 2024, https://doi.org/10.1109/TNNLS.2023.3306421.
143. L. Liu, Y. He, and J. Mao, "Flexible Job-shop Scheduling Method Based on Deep Reinforcement Learning," in *2023 China Automation Congress (CAC)*, Chongqing, China, November 17–19 2023, 2023, pp. 5041–5046.
144. L. Ding, Z. Guan, M. Rauf, and L. Yue, "Multi-policy deep reinforcement learning for multi-objective multiplicity flexible job shop scheduling," *Swarm and Evolutionary Computation*, vol. 87, p. 101550, 2024.
145. D. Rangel-Martinez and L. A. Ricardez-Sandoval, "A recurrent reinforcement learning strategy for optimal scheduling of partially observable job-shop and flow-shop batch chemical plants under uncertainty," *Computers & Chemical Engineering*, vol. 188, p. 108748, 2024, https://doi.org/10.1016/j.compchemeng.2024.108748.
146. C. Su, C. Zhang, C. Wang, W. Cen, G. Chen, and L. Xie, "Fast Pareto set approximation for multi-objective flexible job shop scheduling via parallel preference-conditioned graph reinforcement learning," *Swarm and Evolutionary Computation*, vol. 88, p. 101605, 2024, https://doi.org/10.1016/j.swevo.2024.101605.
147. A. Han and J. J. Yang, "A deep reinforcement learning based solution for flexible job shop scheduling problem," *International Journal of Simulation Modelling*, vol. 20, no. 2, pp. 375–386, 2021, https://doi.org/10.2507/IJSIMM20-2-CO7.
148. L. Wan, L. Fu, C. Li, and K. Li, "Flexible job shop scheduling via deep reinforcement learning with meta-path-based heterogeneous graph neural network," *Knowledge-Based Systems*, vol. 296, p. 111940, 2024, https://doi.org/10.1016/j.knosys.2024.111940.
149. E. Yuan, L. Wang, S. Cheng, S. Song, W. Fan, and Y. Li, "Solving flexible job shop scheduling problems via deep reinforcement learning," *Expert Systems with Applications*, vol. 245, p. 123019, 2024, https://doi.org/10.1016/j.eswa.2023.123019.
150. W. Zhang, F. Zhao, Y. Li, C. Du, X. Feng, and X. Mei, "A novel collaborative agent reinforcement learning framework based on an attention mechanism and disjunctive graph embedding for flexible job shop scheduling problem," *Journal of Manufacturing Systems*, vol. 74, pp. 329–345, 2024, https://doi.org/10.1016/j.jmsy.2024.03.012.
151. J. Tang, Y. Haddad, J. Patsavellas, and K. Salonitis, "Multi-objective Reconfigurable Manufacturing System Scheduling Optimisation: A Deep Reinforcement Learning Approach," *IFAC-PapersOnLine*, vol. 56, no. 2, pp. 11082–11087, 2023, https://doi.org/10.1016/j.ifacol.2023.10.814.
152. J. Popper, V. Yfantis, and M. Ruskowski, "Simultaneous Production and AGV Scheduling using Multi-Agent Deep Reinforcement Learning," *Procedia CIRP*, vol. 104, pp. 1523–1528, 2021, https://doi.org/10.1016/j.procir.2021.11.257.
153. C. Y. Heo, J. Seo, Y. Kim, Y. Kim, and T. Kim, "Estimated Tardiness-Based Reinforcement Learning Solution to Repeatable Job-Shop Scheduling Problems," *Processes*, vol. 13, no. 1, p. 62, 2025, https://doi.org/10.3390/pr13010062.
154. L. Kaven, P. Huke, A. Göppert, and R. H. Schmitt, "Multi agent reinforcement learning for online layout planning and scheduling in flexible assembly systems," *Journal of Intelligent Manufacturing*, vol. 35, no. 8, pp. 3917–3936, 2024, https://doi.org/10.1007/s10845-023-02309-8.

155. C. Shao, Z. Yu, H. Ding, and G. Cao, "A Deep Reinforcement Learning Method for Flexible Job-Shop Scheduling Problem," in *2024 5th International Conference on Electronic Communication and Artificial Intelligence (ICECAI)*, Shenzhen, China, 31.05.2024–02.06.2024, 2024, 184–188, https://doi.org/10.1109/ICECAI62591.2024.10675292.
156. L. Sun, W. Shi, C. Xuan, and Y. Zhang, "Research on Scheduling Algorithm of Knitting Production Workshop Based on Deep Reinforcement Learning," *Machines*, vol. 12, no. 8, 2024, https://doi.org/10.3390/machines12080579.
157. H. Wang, W. Lin, T. Peng, Q. Xiao, and R. Tang, "Multi-agent deep reinforcement learning-based approach for dynamic flexible assembly job shop scheduling with uncertain processing and transport times," *Expert Systems with Applications*, vol. 270, p. 126441, 2025, https://doi.org/10.1016/j.eswa.2025.126441.
158. S. Xu, Y. Li, and Q. Li, "A Deep Reinforcement Learning Method Based on a Transformer Model for the Flexible Job Shop Scheduling Problem," *Electronics (Switzerland)*, vol. 13, no. 18, 2024, https://doi.org/10.3390/electronics13183696.
159. I. Echeverria, M. Murua, and R. Santana, "Solving the flexible job-shop scheduling problem through an enhanced deep reinforcement learning approach," arXiv:2310.15706, Oct. 2023. [Online]. Available: http://arxiv.org/pdf/2310.15706
160. Z. Chen *et al.*, "TBRL: Trajectory-Based Reinforcement Learning for Flexible Job-Shop Scheduling Problem," *Lecture Notes in Computer Science*, no. 15388, pp. 127–142, 2025, https://doi.org/10.1007/978-981-96-0814-0_9.
161. T. E. Thomas, J. Koo, S. Chaterji, and S. Bagchi, "Minerva: A reinforcement learning-based technique for optimal scheduling and bottleneck detection in distributed factory operations," *2018 10th International Conference on Communication Systems and Networks, COMSNETS 2018*, 2018-January, 2018, https://doi.org/10.1109/COMSNETS.2018.8328189.
162. A. Rinciog, C. Mieth, P. M. Scheikl, and A. Meyer, "Sheet-Metal Production Scheduling Using AlphaGo Zero," in *Proceedings of the Conference on Production Systems and Logistics: CPSL 2020*, Stellenbosch, South Africa, March 17–20, 2020, pp. 342–352, https://doi.org/10.15488/9676.
163. S. Luo, "Dynamic scheduling for flexible job shop with new job insertions by deep reinforcement learning," *Applied Soft Computing*, vol. 91, 2020, https://doi.org/10.1016/j.asoc.2020.106208.
164. K. Lei, P. Guo, Y. Wang, J. Zhang, X. Meng, and L. Qian, "Large-Scale Dynamic Scheduling for Flexible Job-Shop With Random Arrivals of New Jobs by Hierarchical Reinforcement Learning," *IEEE Transactions on Industrial Informatics*, pp. 1–12, 2023, https://doi.org/10.1109/TII.2023.3272661.
165. M. Xia, H. Liu, M. Li, and L. Wang, "A dynamic scheduling method with Conv-Dueling and generalized representation based on reinforcement learning," *International Journal of Industrial Engineering Computations*, vol. 14, no. 4, pp. 805–820, 2023, https://doi.org/10.5267/j.ijiec.2023.6.003.
166. J. Chang, D. Yu, Y. Hu, W. He, and H. Yu, "Deep Reinforcement Learning for Dynamic Flexible Job Shop Scheduling with Random Job Arrival," *Processes*, vol. 10, no. 4, 2022, https://doi.org/10.3390/pr10040760.

167. S. Luo, L. X. Zhang, and Y. S. Fan, "Dynamic multi-objective scheduling for flexible job shop by deep reinforcement learning," *Computers & Industrial Engineering*, vol. 159, 2021, https://doi.org/10.1016/j.cie.2021.107489.
168. H. Wang, J. Cheng, C. Liu, Y. Zhang, S. Hu, and L. Chen, "Multi-Objective Reinforcement Learning Framework for Dynamic Flexible Job Shop Scheduling Problem with Uncertain Events," *Applied Soft Computing*, vol. 131, p. 109717, 2022, https://doi.org/10.1016/j.asoc.2022.109717.
169. S. Luo, L. Zhang, and Y. Fan, "Real-Time Scheduling for Dynamic Partial-No-Wait Multiobjective Flexible Job Shop by Deep Reinforcement Learning," *IEEE Transactions on Automation Science and Engineering*, pp. 1–19, 2021, https://doi.org/10.1109/TASE.2021.3104716.
170. T. Zhao, Y. Wang, Y. Tan, and J. Zhang, "Dynamic flexible job shop scheduling algorithm based on deep reinforcement learning," in *2023 35th Chinese Control and Decision Conference (CCDC)*, Yichang, China, May 20–22, 2023, pp. 5099–5104.
171. Y. Zhou, T. Zheng, M. Hu, J. Zhang, and M. Yuan, "Deep reinforcement learning for large-scale flexible Job-shop dynamic scheduling problem," in *IEEE Smart World Congress (IEEE SWC2023) on Automation & Computing*, Portsmouth, UK, August 28–31, 2023, pp. 879–884, https://doi.org/10.1109/SWC57546.2023.10449282.
172. Y. Gui, Z. Zhang, D. Tang, H. Zhu, and Y. Zhang, "Collaborative dynamic scheduling in a self-organizing manufacturing system using multi-agent reinforcement learning," *Advanced Engineering Informatics*, vol. 62, p. 102646, 2024.
173. L. Zhang, Y. Yan, and Y. Hu, "Dynamic flexible scheduling with transportation constraints by multi-agent reinforcement learning," *Engineering Applications of Artificial Intelligence*, vol. 134, p. 108699, 2024, https://doi.org/10.1016/j.engappai.2024.108699.
174. M. Zhang, L. Wang, F. Qiu, and X. Liu, "Dynamic scheduling for flexible job shop with insufficient transportation resources via graph neural network and deep reinforcement learning," *Computers & Industrial Engineering*, vol. 186, p. 109718, 2023, https://doi.org/10.1016/j.cie.2023.109718.
175. Y. Zhang, H. Zhu, D. Tang, T. Zhou, and Y. Gui, "Dynamic job shop scheduling based on deep reinforcement learning for multi-agent manufacturing systems," *Robotics and Computer-Integrated Manufacturing*, vol. 78, p. 102412, 2022, https://doi.org/10.1016/j.rcim.2022.102412.
176. X. Jing, X. Yao, M. Liu, and J. Zhou, "Multi-agent reinforcement learning based on graph convolutional network for flexible job shop scheduling," *Journal of Intelligent Manufacturing*, vol. 35, no. 1, pp. 75–93, 2024, https://doi.org/10.1007/s10845-022-02037-5.
177. Y. Li, W. Gu, M. Yuan, and Y. Tang, "Real-time data-driven dynamic scheduling for flexible job shop with insufficient transportation resources using hybrid deep Q network," *Robotics and Computer-Integrated Manufacturing*, vol. 74, p. 102283, 2022, https://doi.org/10.1016/j.rcim.2021.102283.
178. Z. Wu, H. Fan, Y. Sun, and M. Peng, "Efficient Multi-Objective Optimization on Dynamic Flexible Job Shop Scheduling Using Deep Reinforcement Learning Approach," *Processes*, vol. 11, no. 7, p. 2018, 2023, https://doi.org/10.3390/pr11072018.
179. R. Ma and F. Li, "Knowledge-Specific Reinforcement Learning for Job-Shop Scheduling with Dynamic Processing States in Disassembly Factory," *Communications in*

Computer and Information Science, 2139 CCIS, pp. 160–169, 2024, https://doi.org/10.1007/978-981-97-3948-6_16.

180. L. Wan, X. Cui, H. Zhao, L. Fu, and C. Li, "A novel method for solving dynamic flexible job-shop scheduling problem via DIFFormer and deep reinforcement learning," *Computers and Industrial Engineering*, vol. 198, 2024, https://doi.org/10.1016/j.cie.2024.110688.
181. L. Zhang *et al.*, "Deep reinforcement learning for dynamic flexible job shop scheduling problem considering variable processing times," *Journal of Manufacturing Systems*, vol. 71, pp. 257–273, 2023, https://doi.org/10.1016/j.jmsy.2023.09.009.
182. J. Oren *et al.*, "SOLO: Search Online, Learn Offline for Combinatorial Optimization Problems," in *Proceedings of the Fourteenth International Symposium on Combinatorial Search*, online, July 26–30 2021, 2021, pp. 97–105. [Online]. Available: https://arxiv.org/pdf/2104.01646
183. M. Wang, P. Zhang, J. Zhang, G. Zhang, J. Bi, and M. Jin, "Adaptive Deep Reinforcement Learning for Dynamic Parallel Machines Scheduling with Maintenance Activities," in *2023 28th International Conference on Automation and Computing (ICAC)*, Birmingham, United Kingdom, 30.08.2023–01.09.2023, 2023, pp. 1–6, https://doi.org/10.1109/ICAC57885.2023.10275156.
184. F. Li, S. Lang, B. Hong, and T. Reggelin, "A two-stage RNN-based deep reinforcement learning approach for solving the parallel machine scheduling problem with due dates and family setups," *Journal of Intelligent Manufacturing*, vol. 35, pp. 1107–1140, 2023, https://doi.org/10.1007/s10845-023-02094-4.
185. B. Paeng, I.-B. Park, and J. Park, "Deep Reinforcement Learning for Minimizing Tardiness in Parallel Machine Scheduling With Sequence Dependent Family Setups," *IEEE Access*, vol. 9, pp. 101390–101401, 2021, https://doi.org/10.1109/ACCESS.2021.3097254.
186. M. Sun, J. Ding, Z. Zhao, J. Chen, G. Q. Huang, and L. Wang, "Out-of-order execution enabled deep reinforcement learning for dynamic additive manufacturing scheduling," *Robotics and Computer-Integrated Manufacturing*, vol. 91, p. 102841, 2025, https://doi.org/10.1016/j.rcim.2024.102841.
187. F. Li *et al.*, "A transformer-based deep reinforcement learning approach for dynamic parallel machine scheduling problem with family setups," *Journal of Intelligent Manufacturing*, 2024, https://doi.org/10.1007/s10845-024-02470-8.
188. M. Wang, J. Zhang, P. Zhang, W. Xiang, M. Jin, and H. Li, "Generative deep reinforcement learning method for dynamic parallel machines scheduling with adaptive maintenance activities," *Journal of Manufacturing Systems*, vol. 77, pp. 946–961, 2024, https://doi.org/10.1016/j.jmsy.2024.11.004.
189. S. Lee, J. Kim, G. Wi, Y. Won, Y. Eun, and K.-J. Park, "Deep Reinforcement Learning-Driven Scheduling in Multijob Serial Lines: A Case Study in Automotive Parts Assembly," *IEEE Transactions on Industrial Informatics*, pp. 1–12, 2023, https://doi.org/10.1109/TII.2023.3292538.
190. X. Wang, L. Zhang, T. Lin, C. Zhao, K. Wang, and Z. Chen, "Solving job scheduling problems in a resource preemption environment with multi-agent reinforcement learning," *Robotics and Computer-Integrated Manufacturing*, vol. 77, 2022, https://doi.org/10.1016/j.rcim.2022.102324.

191. S. Wang, J. Li, and Y. Luo, "Smart Scheduling for Flexible and Hybrid Production with Multi-Agent Deep Reinforcement Learning," in *2021 IEEE 2nd International Conference on Information Technology, Big Data and Artificial Intelligence (ICIBA)*, Chongqing, China, December 17–19, 2021, pp. 288–294, https://doi.org/10.1109/ICIBA52610.2021.9688235.
192. D. Shi, W. Fan, Y. Xiao, T. Lin, and C. Xing, "Intelligent scheduling of discrete automated production line via deep reinforcement learning," *International Journal of Production Research*, vol. 58, no. 11, pp. 3362–3380, 2020, https://doi.org/10.1080/00207543.2020.1717008.
193. L. Zhou, L. Zhang, and B. K. Horn, "Deep reinforcement learning-based dynamic scheduling in smart manufacturing," *Procedia CIRP*, vol. 93, pp. 383–388, 2020, https://doi.org/10.1016/j.procir.2020.05.163.
194. Q. Hao and Y. Lv, "A Digital Twin-Based Production-Maintenance Joint Scheduling Framework with Reinforcement Learning," in *2023 8th International Conference on Control and Robotics Engineering (ICCRE)*, Niigata, Japan, April 21–23, 2023, pp. 51–56, https://doi.org/10.1109/ICCRE57112.2023.10155592.
195. M. Imran, G. Antonucci, A. Di Giorgio, F. D. Priscoli, A. Tortorelli, and F. Liberati, "Task Scheduling in Assembly Lines with Single-Agent Deep Reinforcement Learning," in *2023 9th International Conference on Control, Decision and Information Technologies (CoDIT)*, Rome, Italy, July 03–06, 2023, pp. 1583–1588, https://doi.org/10.1109/CoDIT58514.2023.10284428.
196. J. Liu, F. Qiao, M. Zou, J. Zinn, Y. Ma, and B. Vogel-Heuser, "Dynamic scheduling for semiconductor manufacturing systems with uncertainties using convolutional neural networks and reinforcement learning," *Complex and Intelligent Systems*, vol. 8, no. 6, pp. 4641–4662, 2022, https://doi.org/10.1007/s40747-022-00844-0.
197. J. Huang, "Mixed-batch scheduling to minimize total tardiness using deep reinforcement learning," *Applied Soft Computing*, vol. 160, p. 111699, 2024, https://doi.org/10.1016/j.asoc.2024.111699.
198. J. Zhou, L. Zheng, and W. Fan, "Multirobot collaborative task dynamic scheduling based on multiagent reinforcement learning with heuristic graph convolution considering robot service performance," *Journal of Manufacturing Systems*, vol. 72, pp. 122–141, 2024, https://doi.org/10.1016/j.jmsy.2023.11.010.
199. X. Chen and Y. Tian, "Learning to Perform Local Rewriting for Combinatorial Optimization," in *Proceedings of the 33rd International Conference on Neural Information Processing Systems*, Vancouver, Canada, December 08–14 2019, 2019, pp. 6281–6292. [Online]. Available: https://arxiv.org/pdf/1810.00337
200. D. Gankin, S. Mayer, J. Zinn, B. Vogel-Heuser, and C. Endisch, "Modular Production Control with Multi-Agent Deep Q-Learning," in *2021 26th IEEE International Conference on Emerging Technologies and Factory Automation*, Vasteras, Sweden, September 7–9, 2021, pp. 1–8, https://doi.org/10.1109/ETFA45728.2021.9613177.
201. D. Herberger and M. Hübner, Eds., *Proceedings of the Conference on Production Systems and Logistics: CPSL 2024*: Hannover : publish-Ing, 2024.
202. A. Globerson, Ed., *Advances in Neural Information Processing Systems 37 (NeurIPS 2023)*, 2023.
203. M. Medan and R. Cardell-Oliver, *core.edu.au.* [Online]. Available: https://www.core.edu.au/ (accessed: May 10 2025).

204. P. Tassel, M. Gebser, and K. Schekotihin, "An End-to-End Reinforcement Learning Approach for Job-Shop Scheduling Problems Based on Constraint Programming," in *Proceedings of the International Conference on Production Systems and Logistics: CPSL 2021*, Digital Event, August 10–11, 2021, pp. 614–622.
205. S. Huang and S. Ontañón, "A Closer Look at Invalid Action Masking in Policy Gradient Algorithms," in *Proceedings of the Thirty-Fifth International Florida Artificial Intelligence Research Society Conference*, Jensen Beach, Florida, USA, May 15–18, 2022, https://doi.org/10.48550/arXiv.2006.14171.
206. P. A. Alexander, "Domain Knowledge: Evolving Themes and Emerging Concerns," *Educational Psychologist*, vol. 27, no. 1, pp. 33–51, 1992, https://doi.org/10.1207/s15326985ep2701_4.
207. Y. Wang, Q. Zhao, and D. Zheng, "Bottlenecks in production networks: An overview," *Journal of Systems Science and Systems Engineering*, vol. 14, no. 3, pp. 347–363, 2005, https://doi.org/10.1007/s11518-006-0198-3.
208. E. M. Goldratt and J. Cox, *The goal: A process of ongoing improvement,* 30th ed. Great Barrington, MA: The North River Press Publishing Corporation, 2014.
209. C. Waubert de Puiseau, L. Zey, M. Demir, H. Tercan, and T. Meisen, "On The Effectiveness Of Bottleneck Information For Solving Job Shop Scheduling Problems Using Deep Reinforcement Learning," in *Proceedings of the Conference on Production Systems and Logistics: CPSL 2023-2*, Stellenbosch, South Africa, November 14–17, 2023, pp. 738–749, https://doi.org/10.15488/15306.
210. P. Pollett, "Modelling congestion in closed queueing networks," *International Transactions in Operational Research*, vol. 7, 4–5, pp. 319–330, 2000, https://doi.org/10.1016/S0969-6016(00)00004-6.
211. A. M. Law and W. D. Kelton, *Simulation modeling and analysis,* 2nd ed. New York: Mc Graw-Hill, 1991.
212. R. Zhang and C. Wu, "Bottleneck Machine Identification for Shop Scheduling Problems," in *2008 3rd International Conference on Innovative Computing Information and Control*, Dalian, China, 18.06.2008–20.06.2008, 2008, p. 149, https://doi.org/10.1109/ICICIC.2008.186.
213. R. Zhang and C. Wu, "A simulation-based approach to job shop scheduling with bottlenecks," in *2008 6th IEEE International Conference on Industrial Informatics*, Daejeon, Korea (South), 13.07.2008–16.07.2008, 2008, pp. 1007–1012, https://doi.org/10.1109/INDIN.2008.4618250.
214. R. Zhang and C. Wu, "An effective immune particle swarm optimization algorithm for scheduling job shops," in *2008 3rd IEEE Conference on Industrial Electronics and Applications*, 2008, pp. 758–763, https://doi.org/10.1109/ICIEA.2008.4582617.
215. S. J. Mason, J. W. Fowler, and W. Matthew Carlyle, "A modified shifting bottleneck heuristic for minimizing total weighted tardiness in complex job shops," *Journal of Scheduling*, vol. 5, no. 3, pp. 247–262, 2002, https://doi.org/10.1002/jos.102.
216. S. Bär, *Generic Multi-Agent Reinforcement Learning Approach for Flexible Job-Shop Scheduling,* 1st ed.: Springer Vieweg Wiesbaden, 2022.
217. S. Hubert, J. Meintschel, D. Bleidorn, Y. Ortmanns, and R. Wallrath, "Production Scheduling Using Deep Reinforcement Learning and Discrete Event Simulation," *Chemie-Ingenieur-Technik*, vol. 95, no. 7, pp. 1003–1011, 2023, https://doi.org/10.1002/cite.202200242.

218. B. Y. Ekren and B. Arslan, “A reinforcement learning approach for transaction scheduling in a shuttle-based storage and retrieval system,” *International Transactions in Operational Research*, vol. 31, no. 1, pp. 274–295, 2024, https://doi.org/10.1111/itor.13135.
219. L. Guo, Z. Zhuang, Z. Huang, and W. Qin, “Optimization of dynamic multi-objective non-identical parallel machine scheduling with multi-stage reinforcement learning,” in *2020 IEEE 16th International Conference on Automation Science and Engineering (CASE)*, Hong Kong, China, 20.08.2020–21.08.2020, 2020, pp. 1215–1219, https://doi.org/10.1109/CASE48305.2020.9216743.
220. Y. R. Shiue, K. C. Lee, and C. T. Su, “A Reinforcement Learning Approach to Dynamic Scheduling in a Product-Mix Flexibility Environment,” *IEEE Access*, vol. 8, pp. 106542–106553, 2020, https://doi.org/10.1109/ACCESS.2020.3000781.
221. L. M. Steinbacher, A. Ait-Alla, D. Rippel, T. Düe, and M. Freitag, “Modelling Framework for Reinforcement Learning based Scheduling Applications,” *IFAC-PapersOnLine*, vol. 55, no. 10, 2022, https://doi.org/10.1016/j.ifacol.2022.09.369.
222. J. H. Woo, Y. in Cho, S. H. Nam, and J.-H. Nam, “Development of a Reinforcement Learning-Based Adaptive Scheduling Algorithm for Block Assembly Production Line,” in *2021 Winter Simulation Conference (WSC)*, Phoenix, AZ, USA, 12.12.2021–15.12.2021, 2021, pp. 1–12, https://doi.org/10.1109/WSC52266.2021.9715509.
223. C. Waubert de Puiseau, J. Peters, C. Dörpelkus, H. Tercan, and T. Meisen, “schlably: A Python framework for deep reinforcement learning based scheduling experiments,” *SoftwareX*, vol. 22, p. 101383, 2023, https://doi.org/10.1016/j.softx.2023.101383.
224. D. Han, B. Mulyana, V. Stankovic, and S. Cheng, “A Survey on Deep Reinforcement Learning Algorithms for Robotic Manipulation,” *Sensors*, vol. 23, no. 7, 2023, https://doi.org/10.3390/s23073762.
225. Y. Bengio, J. Louradour, R. Collobert, and J. Weston, “Curriculum learning,” in *Proceedings of the 26th Annual International Conference on Machine Learning*, Montreal, Quebec, Canada, 14 06 2009 18 06 2009, 2009, pp. 41–48, https://doi.org/10.1145/1553374.1553380.
226. P. Soviany, R. T. Ionescu, P. Rota, and N. Sebe, “Curriculum Learning: A Survey,” *International Journal of Computer Vision*, vol. 130, no. 6, pp. 1526–1565, 2022, https://doi.org/10.1007/s11263-022-01611-x.
227. S. Narvekar, B. Peng, M. Leonetti, J. Sinapov, M. Taylor E., and P. Stone, “Curriculum learning for reinforcement learning domains: a framework and survey,” *The Journal of Machine Learning Research*, no. 21, pp. 7382–7431, 2022, https://doi.org/10.5555/3455716.3455897.
228. F. Zhuang *et al.*, “A Comprehensive Survey on Transfer Learning,” arXiv:1911.02685, Nov. 2019. [Online]. Available: http://arxiv.org/pdf/1911.02685
229. Y. Bengio, “Evolving Culture Versus Local Minima,” in *Studies in Computational Intelligence*, Vol. 557, *Growing adaptive machines: Combining development and learning in artificial neural networks*, T. Kowaliw, N. Bredeche, and R. Doursat, Eds., Heidelberg, Berlin: Springer, 2014, pp. 109–138.
230. M. Lisicki, A. Afkanpour, and G. W. Taylor, “Evaluating Curriculum Learning Strategies in Neural Combinatorial Optimization,” Workshop on Learning Meets Combinatorial Algorithms at NeurIPS 2020, Nov. 2020. [Online]. Available: https://arxiv.org/pdf/2011.06188

231. A. Nabli and M. Carvalho, "Curriculum learning for multilevel budgeted combinatorial problems," in *Advances in Neural Information Processing Systems 33: Annual Conference on Neural Information Processing Systems 2020*, Virtual, December 6–12, 2020, pp. 7044–7056.
232. R. Zhou, Z. He, Y. Tian, Y. Wu, and S. S. Du, "Understanding curriculum learning in policy optimization for online combinatorial optimization," *Transaction on Machine Learning Research*, 2023. [Online]. Available: https://openreview.net/forum?id=gkebbkruja
233. X. Zhang and G.-Y. Zhu, "A literature review of reinforcement learning methods applied to job-shop scheduling problems," *Computers & Operations Research*, vol. 175, p. 106929, 2024, https://doi.org/10.1016/j.cor.2024.106929.
234. M. S. A. Hameed and A. Schwung, "Reinforcement Learning on Job Shop Scheduling Problems Using Graph Networks," arXiv:2009.03836, Sep. 2020. [Online]. Available: https://arxiv.org/pdf/2009.03836
235. K. Xu, W. Hu, J. Leskovec, and S. Jegelka, "How Powerful are Graph Neural Networks?," in *7th International Conference on Learning Representations*, New Orleans, LA, USA, May 6–9, 2019.
236. A. Kumar and R. Dimitrakopoulos, "Production scheduling in industrial mining complexes with incoming new information using tree search and deep reinforcement learning," *Applied Soft Computing*, vol. 110, 2021, https://doi.org/10.1016/j.asoc.2021.107644.
237. L. Chen *et al.*, "Decision Transformer: Reinforcement Learning via Sequence Modeling," in *Advances in Neural Information Processing Systems 34: Annual Conference on Neural Information Processing Systems 2021*, virtual, December 6–14, 2021, pp. 15084–15097.
238. C. Waubert de Puiseau, C. Dörpelkus, J. Peters, H. Tercan, and T. Meisen, "Beyond Training: Optimizing Reinforcement Learning Based Job Shop Scheduling Through Adaptive Action Sampling," in *Bridging the Gap Between AI Planning and Reinforcement Learning Workshop at the International Conference on Automated Planning and Scheduling 2024*, Banff, Canada, June 2–3, 2024, pp. 1–8.
239. G. E. Uhlenbeck and L. S. Ornstein, "On the Theory of the Brownian Motion," *Physical Review*, vol. 36, no. 5, pp. 823–841, 1930, https://doi.org/10.1103/PhysRev.36.823.
240. D. Pathak, P. Agrawal, A. A. Efros, and T. Darrell, "Curiosity-driven Exploration by Self-supervised Prediction," in *ICML'17: Proceedings of the 34th International Conference on Machine Learning*, Sydney, Australia, pp. 2778–2787.
241. L. Orseau, L. H. S. Lelis, T. Lattimore, and T. Weber, "Single-Agent Policy Tree Search With Guarantees," in *Advances in Neural Information Processing Systems 31: Annual Conference on Neural Information Processing Systems 2018*, Montréal, Canada, December 3–8, 2018, 3205–3215. [Online]. Available: https://arxiv.org/pdf/1811.10928
242. Y. Shen, J. Chen, P.-S. Huang, Y. Guo, and J. Gao, "M-Walk: Learning to Walk over Graphs using Monte Carlo Tree Search," in *Advances in Neural Information Processing Systems 31: Annual Conference on Neural Information Processing Systems 2018*, Montréal, Canada, December 3–8, 2018, pp. 6787–6798.

243. Q. Wang, Y. Hao, and J. Cao, "Learning to traverse over graphs with a Monte Carlo tree search-based self-play framework," *Engineering Applications of Artificial Intelligence*, vol. 105, p. 104422, 2021, https://doi.org/10.1016/j.engappai.2021.104422.
244. W. Kool, H. van Hoof, and M. Welling, "Attention, Learn to Solve Routing Problems!," in *7th International Conference on Learning Representations*, New Orleans, LA, USA, May 6–9, 2019, pp. 1–24. [Online]. Available: https://openreview.net/forum?id=ByxBFsRqYm
245. Y.-D. Kwon, J. Choo, B. Kim, I. Yoon, Y. Gwon, and S. Min, "Pomo: Policy optimization with multiple optima for reinforcement learning," in *Advances in Neural Information Processing Systems 33: Annual Conference on Neural Information Processing Systems 2020*, Virtual, December 6–12, 2020, pp. 21188–21198. [Online]. Available: https://arxiv.org/pdf/2010.16011.pdf
246. A. Holtzman, J. Buys, Du Li, M. Forbes, and Y. Choi, "The Curious Case of Neural Text Degeneration," in *8th International Conference on Learning Representations*, Addis Ababa, Ethiopia, April 26–30, 2020. [Online]. Available: http://arxiv.org/pdf/1904.09751.pdf
247. B. T. Lowerre, "The harpy speech recognition system," Carnegie Mellon University, Pittsburgh/Pennsylvania, 1976.
248. P. Tassel, B. Kovács, M. Gebser, K. Schekotihin, W. Kohlenbrein, and P. Schrott-Kostwein, "Reinforcement Learning of Dispatching Strategies for Large-Scale Industrial Scheduling," in *Proceedings of the International Conference on Automated Planning and Scheduling*, online, June 13–24 2022, 2022, pp. 638–646. [Online]. Available: https://ojs.aaai.org/index.php/ICAPS/article/view/19852#, https://doi.org/10.1609/icaps.v32i1.19852.
249. C. Zhang, W. Song, Z. Cao, J. Zhang, P. S. Tan, and C. Xu, *Learning to Dispatch for Job Shop Scheduling via Deep Reinforcement Learning*. [Online]. Available: https://github.com/zcaicaros/L2D (accessed: Nov. 9 2023).
250. O. Vinyals, S. Bengio, and M. Kudlur, "Order Matters: Sequence to sequence for sets," in *4th International Conference on Learning Representations*, San Juan, Puerto Rico, May 2–4, 2016, pp. 1–11. [Online]. Available: https://arxiv.org/pdf/1511.06391
251. A. Vaswani *et al.*, "Attention Is All You Need," *Proceedings of the 31st International Conference on Neural Information Processing Systems*, pp. 6000–6010, 2017. [Online]. Available: http://arxiv.org/pdf/1706.03762
252. C. Waubert de Puiseau, F. Wolz, M. Montag, J. Peters, H. Tercan, and T. Meisen, "Applying Decision Transformers to Enhance Neural Local Search on the Job Shop Scheduling Problem," *AI*, vol. 6, no. 3, p. 48, 2025, https://doi.org/10.3390/ai6030048.
253. R. Figueiredo Prudencio, M. R. O. A. Maximo, and E. L. Colombini, "A Survey on Offline Reinforcement Learning: Taxonomy, Review, and Open Problems," *IEEE Transactions on Neural Networks and Learning Systems*, vol. 35, no. 8, pp. 10237–10257, 2024, https://doi.org/10.1109/TNNLS.2023.3250269.
254. A. Kumar, A. Zhou, G. Tucker, and S. Levine, "Conservative Q-Learning for Offline Reinforcement Learning," in *Advances in Neural Information Processing Systems 33: Annual Conference on Neural Information Processing Systems 2020*, Virtual, December 6–12, 2020.

255. I. Kostrikov, A. Nair, and S. Levine, "Offline Reinforcement Learning with Implicit Q-Learning," in *The Tenth International Conference on Learning Representations*, online, April 25–29, 2022. [Online]. Available: http://arxiv.org/pdf/2110.06169
256. S. Hu *et al.*, "Q-value Regularized Transformer for Offline Reinforcement Learning," in *Forty-first International Conference on Machine Learning*, Vienna, Austria, July 21–27, 2024, pp. 1–17. [Online]. Available: http://arxiv.org/pdf/2405.17098
257. S. Fujimoto and S. S. Gu, "A Minimalist Approach to Offline Reinforcement Learning," in *Advances in Neural Information Processing Systems 34: Annual Conference on Neural Information Processing Systems 2021*, Virtual, December 6–14, 2021, pp. 20132–20145. [Online]. Available: http://arxiv.org/pdf/2106.06860
258. C. A. Hepburn and G. Montana, "Model-based Trajectory Stitching for Improved Offline Reinforcement Learning," arXiv:2211.11603, Nov. 2022. [Online]. Available: http://arxiv.org/pdf/2211.11603
259. M. Janner, L. Qiyang, and S. Levine, "Offline reinforcement learning as one big sequence modeling problem," in *Advances in Neural Information Processing Systems 34: Annual Conference on Neural Information Processing Systems 2021*, Virtual, December 6–14, 2021, pp. 1273–1286.
260. Z. Zhang, H. Mei, and Y. Xu, "Continuous-Time Decision Transformer for Healthcare Applications," in *Proceedings of The 26th International Conference on Artificial Intelligence and Statistics*, Valencia, Spain, 2023, pp. 6245–6262. [Online]. Available: https://proceedings.mlr.press/v206/zhang23i.html
261. Y. Lai, J. Liu, Z. Tang, B. Wang, J. Hao, and P. Luo, "ChiPFormer: Transferable Chip Placement via Offline Decision Transformer," *International Conference on Machine Learning*, pp. 18346–18364, 2023. [Online]. Available: https://proceedings.mlr.press/v202/lai23c.html
262. W. Ding *et al.*, "Learning to View: Decision Transformers for Active Object Detection," in *2023 IEEE International Conference on Robotics and Automation (ICRA)*, London, United Kingdom, 29.05.2023–02.06.2023, 2023, pp. 1–8, https://doi.org/10.1109/icra48891.2023.10160946.
263. S. Wang, X. Chen, D. Jannach, and L. Yao, "Causal Decision Transformer for Recommender Systems via Offline Reinforcement Learning," in *Proceedings of the 46th International ACM SIGIR Conference on Research and Development in Information Retrieval*, New York, NY, USA, 2023, https://doi.org/10.1145/3539618.3591648.
264. A. Correia and L. A. Alexandre, "Hierarchical Decision Transformer," in *2023 IEEE/RSJ International Conference on Intelligent Robots and Systems (IROS)*, Detroit, MI, USA, 01.10.2023–05.10.2023, 2023, pp. 1661–1666, https://doi.org/10.1109/IROS55552.2023.10342230.
265. S. Hu, L. Shen, Y. Zhang, and D. Tao, "Graph Decision Transformer," in *Advances in Neural Information Processing Systems 36: Annual Conference on Neural Information Processing Systems 2023*, New Orleans, LA, USA, December 10–16, 2023. [Online]. Available: https://arxiv.org/pdf/2303.03747v1.pdf
266. Y.-H. Wu, X. Wang, and M. Hamaya, "Elastic decision transformer," *Advances in Neural Information Processing System 36*, 2024.
267. A. Hussein, M. M. Gaber, E. Elyan, and C. Jayne, "Imitation Learning," *ACM Computing Surveys*, vol. 50, no. 2, pp. 1–35, 2018, https://doi.org/10.1145/3054912.

268. A. Alkhulaifi, F. Alsahli, and I. Ahmad, "Knowledge distillation in deep learning and its applications," *PeerJ. Computer science*, vol. 7, e474, 2021, https://doi.org/10.7717/peerj-cs.474.
269. G. Hinton, O. Vinyals, and J. Dean, "Distilling the Knowledge in a Neural Network," arXiv:1503.02531, Mar. 2015. [Online]. Available: http://arxiv.org/pdf/1503.02531
270. H. Ingimundardottir and T. P. Runarsson, "Discovering dispatching rules from data using imitation learning: A case study for the job-shop problem," *Journal of Scheduling*, vol. 21, no. 4, pp. 413–428, 2018, https://doi.org/10.1007/s10951-017-0534-0.
271. J.-H. Lee and H.-J. Kim, "Imitation Learning for Real-Time Job Shop Scheduling Using Graph-Based Representation," in *2022 Winter Simulation Conference (WSC)*, Singapore, December 11–14 2022, 2022, pp. 3285–3296, https://doi.org/10.1109/WSC57314.2022.10015445.
272. A. Hottung, Y.-D. Kwon, and K. Tierney, "Efficient Active Search for Combinatorial Optimization Problems," in *The Tenth International Conference on Learning Representations*, online, April 25–29, 2022, pp. 1–13. [Online]. Available: http://arxiv.org/pdf/2106.05126
273. I. Bello, H. Pham, Q. Le V, M. Norouzi, and S. Bengio, "Neural Combinatorial Optimization with Reinforcement Learning," arXiv:1611.09940, 2017. [Online]. Available: http://arxiv.org/pdf/1611.09940v3
274. J. Choo *et al.*, "Simulation-guided Beam Search for Neural Combinatorial Optimization," in *Advances in Neural Information Processing Systems 35: Annual Conference on Neural Information Processing Systems 2022*, New Orleans, LA, USA, November 28—December 9, 2022.
275. P. J. M. van Laarhoven, E. H. L. Aarts, and J. K. Lenstra, "Job Shop Scheduling by Simulated Annealing," *Operations Research*, vol. 40, no. 1, pp. 113–125, 1992, https://doi.org/10.1287/opre.40.1.113.
276. E. Nowicki and C. Smutnicki, "A Fast Taboo Search Algorithm for the Job Shop Problem," *Management Science*, vol. 42, no. 6, pp. 797–813, 1996, https://doi.org/10.1287/mnsc.42.6.797.
277. J. Kuhpfahl and C. Bierwirth, "A study on local search neighborhoods for the job shop scheduling problem with total weighted tardiness objective," *Computers & Operations Research*, vol. 66, pp. 44–57, 2016, https://doi.org/10.1016/j.cor.2015.07.011.
278. E. Balas and A. Vazacopoulos, "Guided Local Search with Shifting Bottleneck for Job Shop Scheduling," *Management Science*, vol. 44, no. 2, pp. 262–275, 1998, https://doi.org/10.1287/mnsc.44.2.262.
279. V. Sels, N. Gheysen, and M. Vanhoucke, "A comparison of priority rules for the job shop scheduling problem under different flow time- and tardiness-related objective functions," *International Journal of Production Research*, vol. 50, no. 15, pp. 4255–4270, 2012, https://doi.org/10.1080/00207543.2011.611539.
280. D. P. Kingma and J. Ba, "Adam: A Method for Stochastic Optimization," in *3rd International Conference on Learning Representations*, San Diego, CA, USA, May 7–9, 2015. [Online]. Available: http://arxiv.org/abs/1412.6980
281. J. K. Falkner, „NeuroLS," *GitHub*, 2024. [Online]. Available: https://github.com/jokofa/NeuroLS
282. A. Karpathy, "minGPT," *GitHub*, 2020. [Online]. Available: https://github.com/karpathy/minGPT

283. R. Reijnen, Y. Zhang, Z. Bukhsh, and M. Guzek, "Learning to Adapt Genetic Algorithms for Multi-Objective Flexible Job Shop Scheduling Problems," in *Proceedings of the Companion Conference on Genetic and Evolutionary Computation*, New York, NY, USA, July 15–19, 2023, https://doi.org/10.1145/3583133.3590700.
284. A. Birolini, *Reliability Engineering: Theory and Practice.* Berlin, Heidelberg: Springer, 2004.
285. P. C. Luo, H. Q. Xiong, B. W. Zhang, J. Y. Peng, and Z. F. Xiong, "Multi-resource constrained dynamic workshop scheduling based on proximal policy optimisation," *International Journal of Production Research*, no. 60, pp. 5937–5955, 2022, https://doi.org/10.1080/00207543.2021.1975057.
286. S. Dauzère-Pérès, J. Ding, L. Shen, and K. Tamssaouet, "The flexible job shop scheduling problem: A review," *European Journal of Operational Research*, vol. 314, no. 2, pp. 409–432, 2024, https://doi.org/10.1016/j.ejor.2023.05.017.
287. M. Hutsebaut-Buysse, K. Mets, and S. Latré, "Hierarchical Reinforcement Learning: A Survey and Open Research Challenges," *MAKE*, vol. 4, no. 1, pp. 172–221, 2022, https://doi.org/10.3390/make4010009.
288. N. R. Ke *et al.*, "Modeling the long term future in model-based reinforcement learning," in *7th International Conference on Learning Representations*, New Orleans, LA, USA, May 6–9, 2019. [Online]. Available: https://openreview.net/pdf?id=skgqbn0cf7
289. D. Raposo *et al.*, "Synthetic Returns for Long-Term Credit Assignment," arXiv:2102.12425, Feb. 2021. [Online]. Available: http://arxiv.org/pdf/2102.12425
290. W. Cai, J. Jiang, F. Wang, J. Tang, S. Kim, and J. Huang, "A Survey on Mixture of Experts in Large Language Models," arXiv:2407.06204, 2025. [Online]. Available: https://arxiv.org/abs/2407.06204
291. J. McCormack and M. Grierson, "Building Simulations with Generative Artificial Intelligence," in *Arts, Research, Innovation and Society, Climate Disaster Preparedness: Reimagining Extreme Events through Art and Technology*, D. Del Favero, S. J. Thurow, M. J. Ostwald, and U. Frohne, Eds., Cham: Springer Nature Switzerland; Imprint Springer, 2024, pp. 137–150.
292. D. Palla and A. Slaby, "Evaluation of Generative AI Models in Python Code Generation: A Comparative Study," *IEEE Access*, vol. 13, pp. 65334–65347, 2025, https://doi.org/10.1109/ACCESS.2025.3560244.
293. A. Bou *et al.*, "TorchRL: A data-driven decision-making library for PyTorch," in *The Twelfth International Conference on Learning Representations*, Vienna, Austria, May 7–11, 2024. [Online]. Available: https://openreview.net/forum?id=QxItoEAVMb
294. E. Liang *et al.*, "RLlib: Abstractions for Distributed Reinforcement Learning," in *Proceedings of Machine Learning Research. International Conference on Machine Learning*, Stockholm, Sweden, July 10–15, 2018, pp. 3053–3062.
295. M. Heinicke, *AI-Powered Simulation: Integrating LLMs with Plant Simulation for Next-Gen Models.* [Online]. Available: https://blogs.sw.siemens.com/tecnomatix/ai-llms-siemens-plant-simulation/ (accessed: May 10 2025).
296. M. Faheem, A. Murphy, and C. Reano, "GPU-Accelerated Discrete Event Simulations: Towards Industry 4.0 Manufacturing," in *2021 IEEE Symposium on Computers and Communications (ISCC)*, Athens, Greece, 05.09.2021–08.09.2021, 2021, pp. 1–7, https://doi.org/10.1109/ISCC53001.2021.9631514.

297. B. G. Kang, K.-M. Seo, and T. G. Kim, "Machine learning-based discrete event dynamic surrogate model of communication systems for simulating the command, control, and communication system of systems," *Simulation*, vol. 95, no. 8, pp. 673–691, 2019, https://doi.org/10.1177/0037549718809890.
298. A. Thakur and S. Chakraborty, "A deep learning based surrogate model for stochastic simulators," *Probabilistic Engineering Mechanics*, vol. 68, p. 103248, 2022, https://doi.org/10.1016/j.probengmech.2022.103248.
299. A. Wagenmaker, K. Huang, L. Ke, K. G. Jamieson, Gupta, and Abhishek, "Overcoming the Sim-to-Real Gap: Leveraging Simulation to Learn to Explore for Real-World RL," in *Advances in Neural Information Processing Systems 38: Annual on Neural Information Processing Systems 2024*, Vancouver, BC, Canada, December 10–15, 2024.
300. J. Tobin, R. Fong, A. Ray, J. Schneider, W. Zaremba, and P. Abbeel, "Domain randomization for transferring deep neural networks from simulation to the real world," in *2017 IEEE/RSJ International Conference on Intelligent Robots and Systems (IROS)*, Vancouver, BC, 24.09.2017–28.09.2017, 2017, pp. 23–30, https://doi.org/10.1109/IROS.2017.8202133.
301. D. Smith and K. Simpson, *Functional Safety*: Routledge, 2004.
302. M. Leusin, E. Frazzon, M. Uriona Maldonado, M. Kück, and M. Freitag, "Solving the Job-Shop Scheduling Problem in the Industry 4.0 Era," *Technologies*, vol. 6, no. 4, p. 107, 2018, https://doi.org/10.3390/technologies6040107.
303. V. J. Leon, S. D. Wu, and R. H. Storer, "Robustness Measures and Robust Scheduling for Job Shops," *IIE Transactions*, vol. 26, no. 5, pp. 32–43, 1994.
304. S. Goren and I. Sabuncuoglu, "Robustness and stability measures for scheduling: single-machine environment," *IIE Transactions*, vol. 40, no. 1, pp. 66–83, 2008, https://doi.org/10.1080/07408170701283198.
305. D. Rahmani and M. Heydari, "Robust and stable flow shop scheduling with unexpected arrivals of new jobs and uncertain processing times," *Journal of Manufacturing Systems*, vol. 33, no. 1, pp. 84–92, 2014, https://doi.org/10.1016/j.jmsy.2013.03.004.
306. X.-N. Shen, Y. Han, and J.-Z. Fu, "Robustness measures and robust scheduling for multi-objective stochastic flexible job shop scheduling problems," *Soft Computing*, vol. 21, no. 21, pp. 6531–6554, 2017, https://doi.org/10.1007/s00500-016-2245-4.
307. N. Al-Hinai and T. Y. ElMekkawy, "Robust and stable flexible job shop scheduling with random machine breakdowns using a hybrid genetic algorithm," *International Journal of Production Economics*, vol. 132, no. 2, pp. 279–291, 2011, https://doi.org/10.1016/j.ijpe.2011.04.020.
308. W. Zhu and B. Wang, "New robust single machine scheduling to hedge against processing time uncertainty," in *2017 29th Chinese Control And Decision Conference (CCDC)*, Chongqing, China, 28.05.2017–30.05.2017, 2017, pp. 2418–2423, https://doi.org/10.1109/CCDC.2017.7978920.
309. N. G. Hall and M. E. Posner, "Sensitivity Analysis for Scheduling Problems," *Journal of Scheduling*, vol. 7, no. 1, pp. 49–83, 2004, https://doi.org/10.1023/B:JOSH.0000013055.31639.f6.
310. T. C. E. Cheng and S. Podolsky, *Just-in-time manufacturing: An introduction / T. C. E. Cheng and S. Podolsky,* 2nd ed. London: Chapman & Hall, 1996.

311. W. Gleißner, „Quantitative Verfahren im Risikomanagement : Risikoaggregation, Risikomaße und Performancemaße," *Der Controlling-Berater*, no. 16, 2011.
312. S. Göttlich and S. Knapp, "Uncertainty quantification with risk measures in production planning," *Journal of Mathematics in Industry*, vol. 10, no. 1, 2020, https://doi.org/10.1186/s13362-020-00074-4.
313. Y. Sotskov, N. Y. Sotskova, and F. Werner, "Stability of an optimal schedule in a job shop," *Omega*, vol. 25, no. 4, pp. 397–414, 1997, https://doi.org/10.1016/S0305-0483(97)00012-1.
314. N. Policella, A. Cesta, A. Oddi, and S. Smith, "From precedence constraint posting to partial order schedules: A CSP approach to Robust Scheduling," *AI Communications*, no. 20, pp. 163–180, 2007.
315. P. Kouvelis and G. Yu, *Robust Discrete Optimization and Its Applications*. Boston, MA: Springer US; Imprint: Springer, 1997.
316. J. C. Bean, J. R. Birge, J. Mittenthal, and C. E. Noon, "Matchup Scheduling with Multiple Resources, Release Dates and Disruptions," *Operations Research*, vol. 39, no. 3, pp. 470–483, 1991. [Online]. Available: http://www.jstor.org/stable/171400
317. F. Chollet, "On the Measure of Intelligence," arXiv:1911.01547, Nov. 2019. [Online]. Available: https://arxiv.org/pdf/1911.01547
318. S. C. Y. Chan, S. Fishman, J. Canny, A. Korattikara, and S. Guadarrama, "Measuring the Reliability of Reinforcement Learning Algorithms," in *8th International Conference on Learning Representations*, Addis Ababa, Ethiopia, April 26–30, 2020.
319. A. Ferdowsi, U. Challita, W. Saad, and N. B. Mandayam, "Robust Deep Reinforcement Learning for Security and Safety in Autonomous Vehicle Systems," in *21st International Conference on Intelligent Transportation Systems*, Maui, HI, USA, November 4–7, 2018, pp. 307–312.
320. C. Tessler, Y. Efroni, and S. Mannor, "Action Robust Reinforcement Learning and Applications in Continuous Control," in *Proceedings of the 36th International Conference on Machine Learning*, Long Beach, California, USA, June 9–15, 2019, pp. 6215–6224.
321. M. A. Abdullah *et al.*, "Wasserstein Robust Reinforcement Learning," arXiv:1907.13196, Jul. 2019. [Online]. Available: https://arxiv.org/pdf/
322. D. Mankowitz, T. Mann, P.-L. Bacon, D. Precup, and S. Mannor, "Learning Robust Options," in *Proceedings of the Thirty-Second AAAI Conference on Artificial Intelligence, (AAAI-18), the 30th innovative Applications of Artificial Intelligence (IAAI-18), and the 8th AAAI Symposium on Educational Advances in Artificial Intelligence (EAAI-18)*, New Orleans, Luisiana, USA, February 2–7, 2018, 6409–6416, https://doi.org/10.1609/aaai.v32i1.12115.
323. J. Moos, K. Hansel, H. Abdulsamad, S. Stark, D. Clever, and J. Peters, "Robust Reinforcement Learning: A Review of Foundations and Recent Advances," *MAKE*, vol. 4, no. 1, pp. 276–315, 2022, https://doi.org/10.3390/make4010013.
324. T. Hiraoka, T. Imagawa, T. Mori, T. Onishi, and Y. Tsuruoka, "Learning Robust Options by Conditional Value at Risk Optimization," in *Advances in Neural Information Processing Systems 32: Annual Conference on Neural Information Processing Systems 2019*, Vancouver, BC, Canada, December 8–14, 2019, pp. 2615–2625.

325. W. Dabney, G. Ostrovski, D. Silver, and R. Munos, "Implicit Quantile Networks for Distributional Reinforcement Learning," in *Proceedings of the 35th International Conference on Machine Learning*, Stockholm, Sweden, July 10–15, 2018, pp. 1096–1105. [Online]. Available: http://proceedings.mlr.press/v80/dabney18a.html
326. J. Zhang, A. S. Bedi, M. Wang, and A. Koppel, "Cautious Reinforcement Learning via Distributional Risk in the Dual Domain," *IEEE Journal on Selected Areas in Information Theory*, vol. 2, no. 2, pp. 611–626, 2021, https://doi.org/10.1109/JSAIT.2021.3081108.
327. R. Cheng, G. Orosz, R. M. Murray, and J. W. Burdick, "End-to-End Safe Reinforcement Learning through Barrier Functions for Safety-Critical Continuous Control Tasks," in *The Thirty-Third AAAI Conference on Artificial Intelligence, The Thirty-First Innovative Applications of Artificial Intelligence Conference, The Ninth AAAI Symposium on Educational Advances in Artificial Intelligence*, Honolulu, Hawaii, USA, January 27—February 1, 2019, pp. 3387–3395, https://doi.org/10.1609/aaai.v33i01.33013387.
328. R. Cheng, A. K. Verma, G. Orosz, S. Chaudhuri, Y. Yue, and J. W. Burdick, "Control Regularization for Reduced Variance Reinforcement Learning," in *Proceedings of the 36th International Conference on Machine Learning*, Long Beach, California, USA, 9–15 June, 2019, pp. 1141–1150.
329. Z. Kenton, A. Filos, O. Evans, and Y. Gal, "Generalizing from a few environments in safety-critical reinforcement learning," arXiv:1907.01475v1, Jul. 2019. [Online]. Available: http://arxiv.org/pdf/1907.01475v1
330. A. Hottung, M. Mahajan, and K. Tierney, "PolyNet: Learning Diverse Solution Strategies for Neural Combinatorial Optimization," in *The Thirteenth International Conference on Learning Representations*, Singapore, April 24–28, 2025. [Online]. Available: https://openreview.net/forum?id=TKuYWeFE6S
331. C. F. Hayes *et al.*, "A practical guide to multi-objective reinforcement learning and planning," *Autonomous Agents and Multi-Agent Systems*, vol. 36, no. 1, 2022, https://doi.org/10.1007/s10458-022-09552-y.
332. F. Mischek and N. Musliu, "Preference Explanation and Decision Support for Multi-Objective Real-World Test Laboratory Scheduling," in *Proceedings of the Thirty-Fourth International Conference on Automated Planning and Scheduling*, Banff, Alberta, Canada, June 1–6, 2024, pp. 378–386, https://doi.org/10.1609/icaps.v34i1.31497.
333. A. Hottung, P. Wong-Chung, and K. Tierney, "Neural Deconstruction Search for Vehicle Routing Problems," arXiv:2501.03715, 2025. [Online]. Available: https://arxiv.org/abs/2501.03715
334. L. Ouyang *et al.*, "Training language models to follow instructions with human feedback," in *Advances in Neural Information Processing Systems 35: Annual Conference on Neural Information Processing Systems 2022*, New Orleans, LA, USA, November 28–December 9, 2022.

MIX
Papier aus verantwortungsvollen Quellen
Paper from responsible sources
FSC® C105338

If you have any concerns about our products,
you can contact us on
ProductSafety@springernature.com

In case Publisher is established outside the EU,
the EU authorized representative is:
Springer Nature Customer Service Center GmbH
Europaplatz 3, 69115 Heidelberg, Germany

Printed by Libri Plureos GmbH
in Hamburg, Germany